From Taxonomy to Phylogenetics
Life and Work of Willi Hennig

From Taxonomy to Phylogenetics—Life and Work of Willi Hennig

By

Michael Schmitt

BRILL

LEIDEN • BOSTON
2013

Library of Congress Cataloging-in-Publication Data

Schmitt, Michael, 1949–
From taxonomy to phylogenetics : life and work of Willi Hennig / by Michael Schmitt.
pages cm
Includes bibliographical references and index.
ISBN 978-90-04-21928-1 (hardback : alk. paper) — ISBN 978-90-04-21929-8 (e-book) 1. Hennig, Willi, 1913– 2. Biologists—Germany—Biography. 3. Cladistic analysis. 4. Biology—Classification. I. Title.

QH31.H3585S36 2013
570.92—dc23
[B]

2013003680

This publication has been typeset in the multilingual "Brill" typeface. With over 5,100 characters covering Latin, IPA, Greek, and Cyrillic, this typeface is especially suitable for use in the humanities. For more information, please see www.brill.com/brill-typeface.

ISBN 978-90-04-21928-1 (hardback)
ISBN 978-90-04-21929-8 (e-book)

This book is printed on acid-free paper.

For Wolfgang, Bernd, and Gerd Hennig

CONTENTS

LIST OF ILLUSTRATIONS

Figures

Table

Boxes

ACKNOWLEDGEMENTS

The roots of this book lay in 2008, when Michiel Thijssen from Brill Publishers and I met during the 23rd International Congress of Entomology in Durban (South Africa). While having a chat, I mentioned that Willi Hennig's centenary will be in 2013. Michiel spontaneously suggested that Brill should publish an "authoritative" book on Willi Hennig on that occasion, and invited me to author this book. I had already published some smaller papers on Hennig, and thought I could translate (as most are in German) and combine them, and agreed. For these smaller papers, I used information by many people whom I acknowledged in the respective publications. I want to mention them generally here to emphasise that the present book bases at a considerable degree on their earlier contributions.

Michiel Thijssen, Sabine Steenbeek, and Tessel Jonquière from Brill Publishers always encouraged me, always considered my suggestions, never raised any difficulties for me. I am grateful for this easy co-operation.

Many people and institutions helped me to collect the information, hidden either in public or private archives, or in their memories. Most of all I thank Willi Hennig's three sons, Prof.Dr. Wolfgang Hennig (Kranenburg), Dr. Bernd Hennig (Freiburg im Breisgau), and Gerd Hennig (Tübingen) for providing documents, photographs, verbal information, and critique. Through their help, I could read, copy, and cite piles of up to now unknown documents. They scanned diaries, photographs, notes and whatever they found in their fathers bequest for me. Most of the biographic details given in this book would remain unseen to the public without their support. Moreover, they read and commented on several stages of the manuscript of the present book up to the final.

Prof. Dr. Willi Xylander (Senckenberg Museum für Naturkunde Görlitz) made available documents and photographs from the Willi Hennig-Archive, Görlitz, Ute Kaczinski (Archive of the Senckenberg Deutsches Entomologisches Institut—SDEI, Müncheberg) helped me with documents and photographs from the archive of the SDEI, Walter Pietrusziak (Deutsche Forschungsgemeinschaft—DFG, Bonn) and Undine Beier (Bundesarchiv, Berlin) assisted generously in accessing Willi Hennig's dossier at the DFG, Sabine Strickrodt (Institut für Arbeit und Gesundheit der Deutschen Gesetzlichen Unfallversicherung, Dresden) and Regina Malek (Sächsisches Staatsarchiv, Dresden) made possible to use Fig. 16, from

Prof. Dr. Rolf Kohring (Berlin) I received a photograph of and literature on and by W.G. Kühne. Professor Helmut van Emden (Reading, UK) provided many details of the biography of his father, Fritz van Emden, and checked parts of my manuscript, Prof.Dr. Götz Aly (Berlin) gave many valuable hints regarding my enquiries on information from Germany's NS-era. Maxene Graze (Portland, Oregon, USA) and Gabriele Uhl (Greifswald) critically read the manuscript and improved content and English.

I cordially thank all these and the countless others who helped me.

When I read in the "acknowledgments" sections of other books that the authors gave the names of their family members and thanked them, I often thought that there were an exhibitionistic moment in it. Now, after completing this manuscript, I understand the motives of those colleagues who thanked their family in detail. My wife and our two daughters tolerated my absence from family life the more the longer the writing took, and they not just tolerated it, they made it possible through their constant support and love. Gabriele, Veronika, and Helene—I owe you much more than some words in a book.

Fig. 1. Willi Hennig in February, 1968. Photo taken by “Photostudio Rössle, Ludwigsburg”, the holder of the copyright could not be ascertained.

CHAPTER ONE

INTRODUCTION

Every day is the centenary of millions of people, but only few of these are celebrated by more than some family members. Willi Hennig was born in 1913, and this book is published in 2013 in his memory, to dignify his contributions to science. Clearly, any layperson singled out at random from an average gathering of people on any market square on this planet will hardly know Willi Hennig's name, let aside his scientific merits, other than, e.g., Charles Darwin's. The picture will, however, be completely different if students of the biological discipline of systematics are asked. Of these, hardly any will *not* know at least Hennig's name and one or two terms he coined.

Willi Hennig has dramatically changed the scientific framework of biological systematics. The change was so fundamental, at least it has been perceived as such by quite a number of scientists, that it has been regarded a "scientific revolution". Certainly, appearance and reputation of biological systematics have improved over the past fifty years. The question if *Archaeopteryx* was a member of the stem-group of the modern birds or indeed one of their ancestors, is a serious topic in scientific discourse. Modern authors can treat birds as "Reptilia" without becoming ridiculed. Phrases like "the insects are probably sister-group of a certain subgroup of Crustacea", "the myriapods are probably paraphyletic", or "the complex eye is an apomorph groundplan character of the arthropods" are not taken from an argot of a small circle of adepts. That this is so, is a consequence of the work of Willi Hennig. He introduced a new method of phylogenetic research, which he called 'Phylogenetic Systematics', and which is now in widespread use under the label 'Cladistics'.

Willi Hennig was an unpretentious museum entomologist (Fig. 2) who anyway has shaped the way of reasoning far beyond the limits of insect systematics. The dipterists' community would definitely regard him a luminary even if he had not initiated the development of biological systematics from an art or a handicraft towards a sience in the modern sense. In comparison to his unquestioned merits as a scientific innovator, the public picture of his person is remarkably vague. He was shy and cautious, so he is generally recognised, but rather little is known beyond that about his personality.

Fig. 2. Willi Hennig in his office at the Deutsches Entomologisches Institut, Berlin-Friedrichshagen, in 1960. Courtesy of SDEI, Müncheberg.

The student volunteers assisting at the 26th meeting of the Willi Hennig Society at New Orleans, LA, USA in 2007 wore a badge showing "Willi Hennig Superstar". Likewise, James S. Farris wore a t-shirt showing "Willi Hennig Superstar", when James M. Carpenter took a photo of him in February, 1988 (Hull 1988: 153). What makes a person a "superstar"? How "superstar-like" was Willi Hennig? It is my aim in the present book to trace Willi Hennig's life, his professional career, his scientific work, as detailed as possible. I tried to describe his *curriculum vitae* as exact as possible, but also to understand his feelings and to comprehend his decisions in the light of the contemporary living conditions and his biographical background. He made his way from a small Saxon village to the capital of the German empire, and from there to head of a scientific department, and posthumously to the pedestal of a "superstar".

The present book cannot replace a textbook of Phylogenetic Systematics—for those I refer to Ax (1984) and Wägele (2000), of which English translations exist; shorter and more recent German introductions are Wiesemüller et al. (2003), or Schmitt & Misof (2009). Recommended English readings may be Kitching et al. (1998) or Wiley & Lieberman (2011)—

the list could certainly be longer. Also, I do not claim to present a "history of systematics" or even a "history of cladistics". I refer the reader to Hull's chapters on that topic (1988, 1989), to James S. Farris' special papers (e.g. 1985, 2012), and to the recent treatise by Engel & Kristensen (2013). I focus on Willi Hennig as a person and scientist.

Willi Hennig was no hero, he was a simple man—possibly a "nerd" (Yoon 2009: 240). The more I came to know about his personality, the more likeable I found him. And the more I learned about his scientific work, the more I admired him. I would regret if these attitudes had blurred my scientific sight. Should, however, the readers of this book finally share my view on Willi Hennig's personality and work, I would feel rewarded.

Greifswald, December 15, 2012 Michael Schmitt

Fig. 3. Willi Hennig's place of birth (the right half of the building) in Dürrhennersdorf. Courtesy of Willi Hennig Archive, Görlitz.

CHAPTER TWO

WILLI HENNIG'S BIOGRAPHY

2.1. *A Simple Life in Imperial Germany—*
Willi Hennig's Family Background and Childhood (Fig. 4)

At the dawn of World War I, Marie *Emma* Hennig, née Groß (12.06.1885–03.08.1965), wife of Karl Ernst *Emil* Hennig (28.08.1873–28.12.1947) and living in Dürrhennersdorf near Löbau in Saxony (Upper Lusatia), gave birth on 20.04.1913 to a boy who was named Emil Hans *Willi*. He was born

Fig. 4. The Hennig family in 1916: Emma with Rudolf on her lap, Willi, Emil. Courtesy of Gerd Hennig, Tübingen.

into humble circumstances (Fig. 3). His father had been a simple worker in a quarry and got a job as a railroad worker in 1899; his mother was the illegitimate child of a maidservant who had to care for the pigs at a manor in Nieder-Gebelzig (part of Hohendubrau, Görlitz district, Saxony)(all data on family history from Doc. 001, if not stated otherwise) (Fig. 5). They married on 19.11.1911 in Dürrhennersdorf and had two more sons, Fritz *Rudolf* (05.03.1915–24.11.1990), and Karl *Herbert* (born 24.04.1917, missing since 1943 near Stalingrad) (Figs. 6 and 7).

The earliest genealogical information about an ancestor dates back to 1637, when a peasant by name of Jakob Hennig, aged 63, died in the village of Obercunnersdorf, less than 5 km away from Dürrhennersdorf. As far as there are any data available, all male ancestors on Willi Hennig's father's side were crofters. Emil Hennig was born on 28.08.1873 as the fifth of eight children of Johann Karl *Traugott* Hennig, crofter and churchwarden, and his wife *Johanna* Christiane née Kloß hailed from Sohland bei Rothstein, in Kaana in the parish of Diehsa. This little village was renamed "Reichendorf" in 1936 and vanished with the damming of the lake of Quitzdorf near Niesky in 1975 (Vogel & Xylander 1999). Johanna Hennig died in 1897 on a field at some distance from her house, her husband lived until 1912.

Little is known about his maternal ancestors. Emma Groß's mother Helene *Marie* Groß was born on 18.03.1862 in Weigersdorf and married the

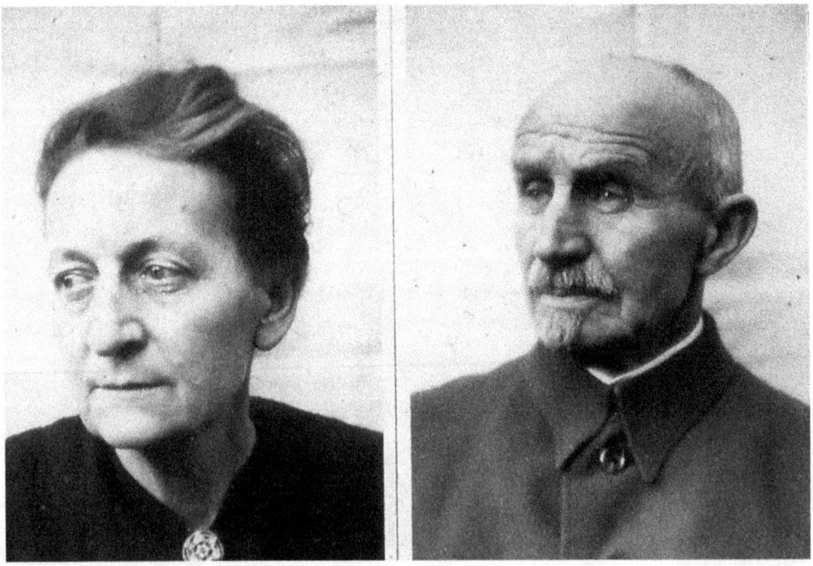

Fig. 5. Emma and Willi Hennig. Courtesy of Willi Hennig Archive, Görlitz.

Fig. 6. The three Hennig brothers: Herbert, Willi, and Rudolf, in 1919. Courtesy of Willi Hennig Archive, Görlitz.

Fig. 7. The Hennig family in 1923: Rudolf, Emil, Herbert, Emma, and Willi. Courtesy of Gerd Hennig, Tübingen.

widowed farm labourer Heinrich Wilhelm Graf in 1898, with whom she lived in Dürrhennersdorf. Emma grew up with her grandmother Johanna Groß née Kaiser (03.04.1836–19.02.1919), widow of Johann Groß (Vogel & Xylander 1999). They lived a miserable life in a thatched clay cottage in Nieder-Gebelzig, from where Emma attended elementary school ("Volksschule") in Ober-Gebelzig, about half a kilometre away, for eight years. She earned some money as a housemaid at the local parsonage and with the leaseholder family by name of Zeisig on Gebelzig's manor. Later she held a modest post in a household in the village of Wilsdruff ca. 15 km east of Dresden. From there she travelled from time to time to Dürrhennersdorf to see her mother. On one of these occasions, she met her husband-to-be Emil Hennig, who by that time was employed as a railroad worker. After marriage she worked in the linen factory of Groß-Schweidnitz, ca. 2 km from her home. This factory was owned by a Scotsman, Sir William Duncan, who possessed a linen merchant business near Leeds in England, but had founded the factory in Saxony because he had exported drapery especially for the German market. Local people did not differentiate between Englishmen and Scotsmen and named the factory "Engelei", a pun that invokes "English" as well as "angel". Accordingly, the ponds built for the water supply to the factory received the nickname "angels' pools" (Engeleiteiche, Doc. 002). When her son Willi was born, Emma ceased working.

When he was just one year old, the First World War broke out. Surprisingly, all he remembered of the war times were several trains transporting troops to the front, which passed through Dürrhennersdorf while the soldiers hung their feet out of the windows. Willi's father was already 41 years old at the start of the war and was not called up for the army. As an employee of the Royal railroad company he was in any case indispensable ("u.k." = unabkömmlich). Rudolf writes of a granduncle (brother of their grandmother) by name of Fritz Kaiser who visited them sometimes, coming from a military hospital at Bautzen, about 25 km from Dürrhennersdorf. It is not clear whether this Fritz Kaiser was at the military hospital as a wounded soldier or for any other reason.

We know nothing about Willi Hennig's thoughts and feelings throughout these early years of his life. At the end of WWI (11.11.1918), he was certainly old enough to understand that something dramatic had happened. The conditions of the Treaty of Versailles (signed June 28, 1919) depressed people all over Germany in those days; it can be taken for granted that also in the village of Dürrhennersdorf the adults were concerned and discussed on this matter in the pubs and at home. However, there is no indication that the Hennig boys—even the two older ones—perceived anything that remained in their memory.

The Hennigs were Protestants, but seemingly were not very enthusiastic about religion. Rudolf Hennig emphasised in his notes (written ca. 1990, when he was a retired parson) that he and his brothers joined Sunday school regularly, but he mentions that Willi often stayed at home reading books. Rudolf also writes that his father had to miss church services many times, due to his duty to check the rails on Sunday when there was less traffic on them. Taking into account that Willi Hennig's widow Irma described her parents-in-law as little religious but politically conservative and patriotic Prussians (Doc. 003), we can safely assume that Willi's mental development was not deeply influenced by religious content.

Willi entered primary school at Dürrhennersdorf April (Easter) of 1919. The schoolhouse was only about 100 m from Willi's home; likewise the little church (Fig. 8). He was an excellent pupil from the beginning. According to his brother Rudolf's report, the teachers realised Willi's aptitude—and diligence—immediately. Therefore, we can imagine that Willi received special attention even in primary school. Willi is described by

Fig. 8. View from Willi Hennig's place of birth to the church of Dürrhennersdorf. Original, taken 20.11.2007.

his brother as behaving quietly and avoiding conflicts whenever possible. He rarely played with his younger brothers, or perhaps never, as Rudolf could not remember even a single instance. At the same time, Willi was not arrogant or a mummy's boy. He simply preferred staying at home and reading a book to joining his brothers' activities.

Only after half a year, in the autumn of 1919, the family moved to Taubenheim (Spree), less than 2 km from the Czech border and 10 km—as the crow flies—from Dürrhennersdorf. Here, Willi attended the primary school until 1921 (Fig. 9). The family lived in a house for signalmen near the local station, together with grandmother Marie Graf (Fig. 10).

Formally, the move from Dürrhennersdorf to Taubenheim was caused by an official transfer of Willi's father. However, Vogel & Xylander (1999) speculate that Emma Hennig's emotional restlessness might also have played a part. Rudolf described his father as having a calming influence on family life, whereas Emma Hennig must have been a difficult character, nervous and unstable. According to Rudolf Hennig, she suffered at times from attacks of depression, and she easily clashed with neighbours wherever she lived. There have been dramatic scenes when mother Hennig locked herself in her bedroom and there produced sounds of

Fig. 9. Willi Hennig as a pupil in Taubenheim/Spree in 1920/21 Taubenheim/Spree, the teacher, Mr. Ritter standing, W.H. 1st row on the Left, at the aisle. Courtesy of Gerd Hennig, Tübingen.

Fig. 10. The house for signalmen in Taubenheim. The Hennig family lived in the right half of the building. Original, taken 20.07.2012.

moaning and crying and left the room only after several days, then behaving as if nothing at all had happened.

Already in November 1921, the Hennig family including grandmother Marie Graf moved again, after Emil Hennig was promoted to squad leader (Rottenführer). They moved over 3 km Northeast to Oppach, where they lived in the right wing of the upper floor of the station building (Fig. 11). As a result, Willi attended the local elementary school from November 11, 1921. As Rudolf reports, here they had electric light for the first time. Before that, they had only petroleum lamps, and during their time in Taubenheim only carbide lamps, which frightened the boys because of the gurgling noise they produced.

During the first years of their time in Oppach, money rapidly lost value in Germany (Fig. 12). One loaf of rye bread cost 23 Mark in October, 1922, 2000 Mark in July, 1923, and 260 million Mark at the peak of the inflation, in November 1923. A ticket for a ride on the Berlin tramway was 50 Mark in early 1923, in July already 1000 Mark, and by mid-November it was 150 billion Mark (Doc. 004)—today, it is € 2.40. Millions of Germans lost their job or otherwise became impecunious. Luckily, Emil Hennig remained employed, and the family had a small piece of land on which to grow vegetables. Thus, although the children realised that there was a shortage of money, they never had to starve and survived this hard time

Fig. 11. The former station building in Oppach. The Hennigs lived in the right half of the upper floor (hidden by the tree). Original, taken 20.07.2012.

Fig. 12. Bill of one billion Mark, originally issued December 15, 1922, for 1000 Mark, overprinted ca. in March, 1923. Original.

remarkably unimpaired. This was primarily due to the commitment of Emma Hennig—the boys had to call her "Mama", whereas they had to address Marie Graf as "Mother"—who sacrificed herself for her children. In spite of her difficult character, she was a loving mother who told her kids self-invented fairy tales and made sure that all members of the family had enough to eat every day. Rudolf Hennig stated (Doc. 001: 13) that their mother would have given her life for her boys without hesitation.

Clearly, Emma Hennig suffered from the social stigma of her illegitimate birth. At any rate, she was very ambitious to provide her sons with an excellent education and school training (Fig. 13). Already during Willi's primary school years in Oppach, he received—together with three older girls—private lessons in French and mathematics (Doc. 005). His mother had engaged their neighbour Dr. Reinhold Seifert, a retired military physician (Oberstabsarzt = chief staff surgeon), who not only taught these subjects to Willi but also directed him to collect insects and build up a herbarium.

Even though the atmosphere at home was not specifically religious, it was entirely natural that Willi was confirmed (Fig. 14). This celebration took place on the 10th of April, 1927. His confirmation verse was Revelations 3:20 (Here I am! I stand at the door and knock. If anyone hears my voice and opens the door, I will come in and eat with that person, and they with me [New International Version]).

After having completed the normal eight-year German elementary school curriculum, he passed the entrance examination for the Landesschule (special type of Highschool) of Klotzsche near Dresden in spring, 1927. His mother had accompanied him when he travelled to Dresden. Willi passed the examination with distinction, certainly due to the intensive extra lessons he had received. When he and his mother returned to Oppach, they were proud, felt relieved, and expected congratulations from the other family members. Instead, however, the two younger boys, Rudolf and Herbert, ran towards them crying miserably, together with their grieving father. Grandmother Marie Graf had died. While cleaning the windows of the flat she stepped onto a chair from which she fell and ruptured a vein. A neighbour, with whom the Hennigs had a difficult relationship (as with all earlier neighbours), had found her and alerted the boys. They ran to call a local physician, since their father was working and would return only later. When the doctor arrived, Marie Graf could hardly speak and died soon afterwards.

Thus, a family tragedy overshadowed Willi Hennig's entry into a crucial new phase of his life. After leaving elementary school for the Gymnasium in Dresden he never returned to his hometown for any length of time.

Fig. 13. The elementary school of Oppach. Original, taken 20.07.2012.

Fig. 14. The protestant church of Oppach, where Willi Hennig received confirmation in 1927. Original, taken 20.07.2012.

2.2. *A Gifted Young Man—*
Time in Dresden, at School and at the Museum (Fig. 15)

Willi Hennig entered the Landeschule Klotzsche near Dresden in April, 1927—after Easter. This was a boarding school that followed the pedagogical ideas of Hermann Lietz. Hermann Lietz was inspired by the concepts

MITTEILUNGEN
AUS DER LANDES-
SCHULE DRESDEN

KLOTZSCHE B./DRESDEN JUNI 1931

SÄGEBOCK-KÄFER

HEFT 8

Fig. 15. Title page of issue no. 8 of the Mitteilungen aus der Landesschule Dresden, in which Willi Hennig's first publication appeared in June, 1931, showing the longicorn beetle *Prionus coriarius*, on which Hennig commented in his contribution. Courtesy of Bernd Hennig, Freiburg im Breisgau.

of Cecil Reddie, the founder of Abbotsholme School. Although the State of Saxony ran the Landesschule, the style of education there was liberal and progressive. Much emphasis was laid on outdoor activities and on social skills such as compassion. The fees were generally low, and more than 50 % of the places were waived. It was an explicit aim of the school to boost the proportion of boys from lower social classes that entered university. The school was founded in 1920 and had moved into a newly erected complex of buildings in Easter, 1927 (Amlung 2009). Thus, Willi Hennig belonged to the first cohort of pupils to use the new rooms (Fischer et al. 1927).

The 250 students lived in twelve groups each together with a—normally married—teacher in one of the six residential houses grouped around a green square (Fig. 16). Willi Hennig lived in the group supervised by a teacher by the name of Willy Matthes in the "Abteilung 11". Willy Matthes taught biology and was a Social Democrat. Seemingly, he was fond of Ernst Haeckel and Ludwig Büchner (Doc. 001: 34) and had a profound influence on Willi Hennig. Rudolf Hennig, who entered the Landesschule two years after Willi, described Mrs. Matthes as a "very liberal protestant". Since the teachers lived with the students in a family-like group, this means that Willi Hennig evidently enjoyed an open minded atmosphere during his

Fig. 16. Aerial view of the Landesschule Dresden, taken in 1927. Courtesy of Sächsisches Staatsarchiv, Dresden.

years at the Landesschule. Mr. Matthes directed Willi Hennig to collecting insects, and Willi helped him to order and rearrange the entire school collection of animals and fossils, which was especially rich in butterflies (Doc. 007).

Willi Hennig wrote a diary, at least in his first year at the Landesschule, of which four pages have survived and are kept by his son Gerd Hennig (Doc. 008). In this document, the fourteen-year-old pupil reports details of his daily routine—May 8, 1927: "today I had table service, tomorrow I shall have room service", remarkable events—May 12, 1927 on the field day two days earlier, tests in French and Latin, and comments on the lunch menu ("today we had rhubarb for the first time"). All in all, we see the writing of a rather naive boy who meticulously describes affairs which some others would not have considered mentioning at all. However, he found a football (soccer) tournament also worth reporting, even if four days late (Fig. 17). He seems to have been an absolutely "normal" pupil. He would worry about the result of a test, he was glad that the team from his school won a football tournament against a local club, he liked spending a night's leave ("Nachturlaub") with his parents and brothers.

The most stunning work of his school days is a composition test of thirty-one pages (including tables, illustrations, and references) on the position of systematics in zoology, dated May 4, 1931 (Doc. 009). He treated in a surprisingly lucid way—at the age of 18—problems as the delineation of species, the use of embryology and fossils in systematics, and the representation of genealogic relationships in the system. This essay gives clear evidence of Willi Hennig's serious interest in the subject, and also demonstrates his profound knowledge of it. He received the maximum score for it ("sehr gut"), although the subject of the test was German, not biology (Schmitt in press). As in earlier years, he showed more interest in natural history and reading science books than in social activities with his peers. Most certainly, this is the reason that he received the nickname "Orang" (after Orang Utan) at the Landesschule (Doc. 001: 41; Doc. 010).

During his last year, the first published article by Willi Hennig appeared, in the school's journal ("Mitteilungen aus der Landesschule Dresden"—see Fig. 15) in 1931. In this paper (no. 168), the young author reports that he had found 80 insect species over the past three years on the area of the Landesschule. He lists 73 of these species along with notes on their ecology, taxonomy, and nomenclature. Thus, it is evident that he had observed insects at least from the age of fifteen on, and that he had developed a special interest also in naming and ordering organisms.

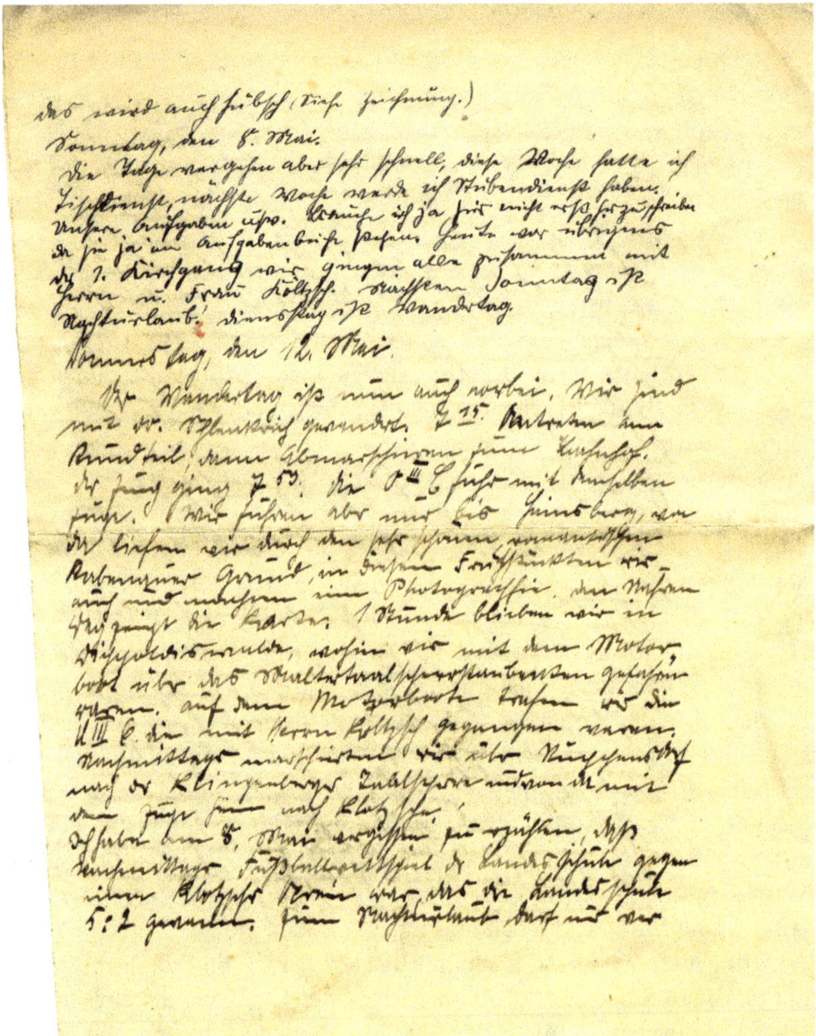

Fig. 17. Entry in Willi Hennig's diary for 08./12.05.1927. Transcript of the handwritten Text:

Das wird auch hübsch (siehe Zeichnung)

Sonntag, den 8. Mai [1927]

Die Tage vergehen aber sehr schnell, diese Woche hatte ich/ Tischdienst, nächste Woche werde ich Stubendienst haben./ Unsere Aufgaben usw. brauche ich ja hier nicht erst hinzuschreiben/ da sie ja im Aufgabenbuch stehen. Heute war übrigens/ der 1. Kirchgang wir gingen alle zusammen mit/ Herrn u. Frau Költzsch. Nächsten Sonntag ist/Nachturlaub, Dienstag ist Wandertag.

Donnerstag, den 12. Mai

Der Wandertag ist nun auch vorbei. Wir sind/ mit Dr. Schenkrich gewandert. 7^{15} Antreten am/ Rundteil, dann Abmarschieren zum Bahnhof./ Der Zug ging 7^{50}. Die O^{III}C fuhr mit demselben/ Zuge. Wir fuhren aber nur bis Hainsberg, von/ da liefen wir durch den sehr schön romantischen/ Rabenauer Grund, in diesem frühstückten wir/ auch und machten eine Photographie. Den näheren/ Weg zeigt die Karte. 1 Stunde blieben wir in/ Dippoldiswalde, wohin wir mit dem Motor/ boot über das Maltertaalscherrstaubecken gefahren/ waren. Auf dem Motorboot trafen wir die/ UIIIB, die mit Herrn Költzsch gegangen waren./ Nachmittags marschierten wir über Ruppendorf/ nach der Klingenberger Taalsperre und von dort mit/ dem Zuge heim nach Klotzsche.

Ich habe am 8. Mai vergessen zu erzählen, daß/ nachmittags Fußballwettspiel der Landesschule gegen/ einen Klotzscher Verein war, das die Landesschule/ 5:7 gewann. Zum Nachturlaub darf nur wer/

[bei 25 lat. Vokabeln nicht mehr als 2 Fehler macht, hoffentlich geht alles gut, es wäre doch dumm, wenn ich erst später fortkönnte.]

Translation:

This also will get nice (see drawing)

Sunday, May 8 [1927]

The days pass really quickly, this week I had/ table service, next week I shall have room service./ I do not need to write down our duties here/ as they are listed in the record of duties. By the way, today was/ the first churchgoing we all went together with/ Mr. and Mrs. Költzsch. Next Sunday is/ night leave, Tuesday is field day.

Thursday, May 12

Now the field day is over, too. We have/ hiked with Dr. Schenkrich. At quarter past seven a.m./ Lining up at the rondel, then marching off for the station./ The train left at ten to eight. The O^{III}C [= Obertertia C, which means the class one grade further, as Willi Hennig was in the U^{III}- Untertertia—at that time] went on the same/ train. But we only went to Hainsberg, from/ there we hiked through the very beautifully romantic "Rabenauer Grund" [a valley], in which we had breakfast/ and where we took a photograph. The closer route shows the map [seemingly, Willi had included a self-drawn sketch of a printed map in this diary]. 1 hour we stayed at/ Dippoldiswalde, whereto we had travelled by motor/ boat over the "Maltertaalscherrstaubecken" [a dammed lake]/ On the motor boat we met/ the UIIIB [a parallel class] who had gone with Mr. Költzsch./ In the afternoon we marched via Ruppendorf/ to the Klingenberger Taalsperre [another dammed lake] and from there by/ train back home to Klotzsche.

On May 8, I had forgotten to tell that/ in that afternoon there was a football [soccer] tournament of our school against/ a club from Klotzsche, which our school won/ 5:7. For the night leave, only those are admitted/

[who have less than three mistakes among 26 Latin vocables. Hopefully everything goes well, it would be too silly if I could leave only later.]

He passed the final examination (Abitur) on February 26, 1932, with good or very good grades in nearly all disciplines. He had, e.g., reached a "very good" ("sehr gut—1") in Natural History, a "good" ("gut—2a") in Music, a very good ("sehr gut—1b") in Latin, a good ("gut—2a") in English, a "good" ("gut—2") in Spanish, and only in Sports had he received only a "sufficient" ("befriedigend—3")(Doc. 011).

In 1930 by the latest, Willi Hennig must have moved to the school group supervised by the lecturer Maximilian Rost, since Wilhelm Meise (see box 1), from 1929 on keeper of all animal taxa except insects at the Dresden State Museum of Natural History, reported (Doc. 006) that the biology teacher Mr. Rost showed up one day at his office, accompanied by Willi Hennig who was in his fourth year at the Landesschule and living in Mr. Rost's group. As Wilhelm Meise remembered the school grade of Willi Hennig ("Obersekunda") well, this fateful encounter must have happened between April and September, 1930. Willi Hennig was able to skip a grade

Box 1 Wilhelm Meise

Wilhelm Meise was born on September 12, 1901 in Essen, Germany. Originally, he intended to become an elementary school teacher, but could not find a position after completing his training in 1921. He earned his living with temporary jobs, e.g. by giving cello lessons and as a tutor in the house of a head forester in Anhalt (Hoerschelmann & Neumann 2003). From 1924 to 1928 he studied Sciences and Mathematics at the University of Berlin, where he received his doctorate (Ph.D.) for his dissertation on the geographical distribution of crows of the *Corvus-corone*-complex, supervised by Erwin Stresemann. In 1929, he was hired as a curator for all animal taxa at the State Museum for Zoology and Anthropology (Staatliches Museum für Tierkunde und Völkerkunde) in Dresden (Fig. 76). In August, 1939, he was called up for the army, was taken prisoner of war in 1945 by the Russians and could not return from Siberia until 1948. Through the help of Erwin Stresemann, he found a provisional position at the Berlin Zoological Museum. In 1951 he was appointed curator of molluscs at the Zoological Museum of Hamburg, where he was able to transfer to the department of ornithology in 1956. He lived in Hamburg after his retirement in 1972 until he died on 24.8.2002, aged 101 (Fig. 77).

Wilhelm Meise was one of the most important and seminal German ornithologists, who authored or co-authored five major handbooks or popular series. He was also interested in all flying (or gliding) vertebrates, such as frogs, snakes, lizards, and bats. Most of his 170 publications dealt with *birds*, but occasionally he also worked and published on scorpions, spiders, lizards, snakes, and molluscs (Haffer 2003; Steinbacher 1972).

Fig. 76. Wilhelm Meise ca. 1935. Courtesy of Wilhelm Meise (1999).

Fig. 77. Wilhelm Meise on 17.02.2000 (Original).

on 29 September, 1930—certainly due to his profound training in addition to his splendid performance. Thus, after that date, he was in his fifth grade ("Unterprima") at the Landesschule. According to Wilhelm Meise, Mr. Rost had reached a point where he could no longer teach anything new in natural history to Willi Hennig. Therefore, he brought Hennig into contact with Meise, who hopefully could provide further training in zoological systematics.

Willi Hennig gave a slightly different account of this temporal sequence. In an application for a grant from the Studienstiftung des Deutschen Volkes (Doc. 006), he wrote that he was acquainted with museum entomology since "OIII" (Obertertia, which would mean since the time between April, 1928 and April, 1929). At that time, he still lived in the group of Mr. Matthes, as documented by his brother Rudolf. Consequently, it should have been Mr. Matthes who brought Willi to the Dresden museum of natural history. As Wilhelm Meise was 97 years old when he wrote his report, but could recall all the details of his time at the Dresden museum when I interviewed him two years later, I have no doubt that his mention of Mr. Rost is correct. Maybe a mistake slipped in when Willi Hennig wrote "OIII" instead of "OII", or he extended the period of his experience with museum entomology by preponing the starting point, in order to enhance his chances to receive the desired grant.

At any rate, Mr. Rost's astounding self-awareness had the consequence that Willi Hennig got into contact with science outside his school. He came to the museum, which was housed until 1937 in the South-East wing ("Pavillon") of the Dresden Zwinger (Fig. 18) (Steinbacher 1972) regularly once a week (Doc. 005: 1). Wilhelm Meise entrusted him with the task of compiling an index to the 4000-page-list of fishes in the British Museum of Natural History, as a kind of test of Hennig's seriousness and perseverance (Doc. 005). Soon, Meise recognised how motivated, gifted, and already well-trained Hennig was, and he invited him to co-author two papers on the "flying" snakes of the genera *Chrysopelea* and *Dendrophis* (Meise & Hennig 1932, 1935).

Even before the encounter with Wilhelm Meise, he had got—mediated through his former teacher, Mr. Matthes (Doc. 006)—into contact with Fritz van Emden, at that time keeper of all insect taxa at the Dresden museum and studying the taxonomy of ground beetles (Carabidae). Wilhelm Meise certainly felt that Willi Hennig did not wish to become an entomologist before he met Fritz van Emden (Doc. 013). Be it as it may be, young Willi took quite a fancy of the six-legged critters and certainly also of Dr. van Emden. As Meise described it (Doc. 005), "van Emden had his office in the South-East pavillon of the Zwinger, above the director's

Fig. 18. Southeast ("Porcelain") pavillon of the Dresden Zwinger. In this building was the entomology collection housed in the 1930s. Original, taken 18.07.2012.

and my office. Mr. Hennig wanted to learn more. He immediately became an entomologist. I saw him sometimes disappear upstairs" ("van Emden arbeitete im Südost-Pavillon des Zwingers über dem Direktor- und meinem Raum. Herr Hennig wollte mehr lernen. Er wurde sogleich Entomologe. Ich sah ihn manchmal nach oben verschwinden").

2.3. *Studying Insects* (Fig. 19)

Certainly stimulated and guided by Fritz van Emden (see box 2), Willi Hennig began morphological and taxonomic investigations on Diptera, which he continued as a university student. In parallel, he completed the papers on "flying" snakes (List of Hennig's publications: nos. 1, 5, 9, 14). He enrolled at the University of Leipzig on April 28, 1932 under the number 1699 for zoology, botany, and geology. His application to the *Studienstiftung des Deutschen Volkes* (Doc. 006) was successful and he received a grant of 1021 Reichsmark (RM) for the period of three semesters—307 RM for the first, 407 RM for the second, and 307 RM for the third semester (Doc. 017). In addition, his parents subsidised him with the amount of 500 RM, which they had to take up as a loan from a bank (Doc. 173: Personal-Fragebogen).

Fig. 19. Photo of Willi Hennig, from his student ID card, issued 28.04.1932. Courtesy of Gerd Hennig, Tübingen.

Box 2 Fritz van Emden

Fritz Isidor van Emden was born on 03.10.1898 in Amsterdam (The Netherlands) but moved to Leipzig (Germany) in 1900 as a child and attended school there. From 1918 to 1921 he studied Zoology at the Leipzig University, receiving his doctorate for a thesis on the reproductive biology of the dipper (*Cinclus cinclus*), supervised by Johannes Meisenheimer. Although originally working on an ornithological subject, he specialised in entomology, above all on the systematics of beetle larvae. After several temporary jobs, amongst them some in applied entomology, he was appointed keeper of insects at the State Museum for Zoology and Anthropology (Staatliches Museum für Tierkunde und Völkerkunde) in Dresden (Fig. 78) on 1.4.1927. According to Hennig (1960), this position was certainly the one Fritz van Emden desired most of all. However, due to the racist laws of the National Socialists in Germany, van Emden was expelled on 30.9.1939 from the museum. In a letter to Wilhelm Meise of 28.07.1948, he explained that he was of partly non-Aryan descent ("teilweise nicht arischer Abstammung", Doc. 012). His son, Helmut van Emden, explained (Doc. 014) that Fritz van Emden's mother was Jewish, thus he was regarded "half-Jew" by the Nazis and consequently dismissed from public service. Luckily, Fritz van Emden and his family could emigrate to London in 1936, where they lived on a grant from the Society for the

Protection of Science and Learning. He found a position at the Imperial (later: Commonwealth) Institute as a taxonomist for calyptrate Diptera, since the rules of the institute forbade staff from working on a taxon for which they had a private collection. He was later put in charge of the collection of Diptera-Cyclorhapha at the British Museum (Natural History). He made no secret of the fact that flies were only a means of livelihood, and that beetles continued to absorb his interest (Oldroyd 1958).

Fritz van Emden's most important contribution to entomology were arguably the seven parts of "Larvae of British Beetles" which appeared from 1939 to 1949. Nearly as valuable are his contributions to the *Handbooks for the Identification of British Insects* and to the *Fauna of India*. According to Hennig (1960), van Emden's papers on the Muscidae and Tachinidae of the Ruwenzori-expedition will for a long time remain the basis for all studies on the Ethiopian species of these two large and difficult Diptera families.

Fritz van Emden died on 02.09.1958 in London.

Fig. 78. Fritz Isidor van Emden ca. 1958 (Courtesy of Helmut van Emden, Reading).

There is no evidence that Willi Hennig was involved in the political and social overthrows of the late 1920s and early 1930s. Nor is there any sign that he cared about. He must have been an avid student who focused entirely on his scientific subjects. Seemingly, he aimed at skipping a basic zoology course, which he probably thought might have been boring after all his experience at the Dresden museum. His fatherly mentor, Fritz van Emden, confirmed in an official letter (Doc. 018) that Willi Hennig had acquired zoological skills and that he was able to work scientifically on a high level, as proven by the paper in press on the "flying" snakes, jointly written with Wilhelm Meise (no. 1). From this letter can be taken that he had planned to switch to Rostock University after his third semester. Possibly, this idea was only brought up for tactical reasons. Amusingly, Fritz van Emden attached a Postcard (Doc. 019) to that letter, in which he gave advice "privatissime" how Willi Hennig should behave towards professor Meisenheimer who had to decide on this request. Fritz van Emden agreed that professor Meisenheimer did not give a kind and open impression. He recommended that Hennig should not appear intimidated but arise in a cool and self-assured manner. He wrote: "when you have succeeded by this method once, you will reach your goals so much easier in the future" ("Wenn Sie diese Maxime einmal mit Erfolg durchgeführt haben, werden Sie in Zukunft umso leichter Ihre Ziele erreichen"). According to Willi Hennig's record of study (Tab. 1), it did not work—he had to attend the

Table 1. List of courses and lectures Willi Hennig attended at Leipzig university (as listed in his record of study (Doc. 015):

Summer semester 1932:

Allgemeine Zoologie (general zoology)—Prof.Dr. [phil.habil. Johannes] Meisenheimer
Allgemeine Botanik (general botany)—Prof.Dr. [phil.habil. Wilhelm] Ruhland
Systematik II Blütenpflanzen (systematics 2 flowering plants)—Prof.Dr. [phil. Fritz] Bachmann
Mikroskopische Übungen für Anfänger (training in microscopy for beginners)—Prof.Dr. Ruhland
Spezielle Zoologie der Insekten (special zoology of insects)—Prof.Dr. [Otto Hermann] Steche
Experimentalchemie I. Teil/Anorganische Chemie (experimental chemistry part 1, inorganic chemistry)—Prof.Dr. [phil., Dr.rer.nat.h.c. Burckhardt] Helferich
Zoologischer Anfängerkurs für Naturwissenschaftler (zoological beginners course for scientists)—Prof.Dr. Meisenheimer
Botanische Exkursionen (botanical excursions)—Prof.Dr. Bachmann and Dr. K. [Karl] Wetzel

Table 1. (*cont.*)

Zoologische Exkursionen (zoological excursions)—Dr. A. [Prof. Dr. phil. Arno] Wetzel

Winter semester 1932/33:

Vergleichende Morphologie und Anatomie der Tiere 5-stündig (comparative animal morphology and anatomy, 5 hours per week)—Prof.Dr. Meisenheimer
Großes zoologisches Praktikum (advanced course in zoology)—Prof.Dr. Meisenheimer
Pflanzenphysiologe (plant physiology)—Prof.Dr. Ruhland
Systematik I—Prof.Dr. Bachmann
Zoologisches Colloquium.

Summer semester 1933:

Spezielle Zoologie der Vögel (special zoology of birds)—[Prof.Dr.phil. *Georg* Johann] Grimpe
Pflanzenphysiologische Übungen (laboratory course in plant physiology)—K. Wetzel
Geologische Formationskunde (geological stratigraphy)—[Prof.Dr.phil. Franz] Kossmat
Großes zoologisches Praktikum (advanced course in zoology)—[Prof. Dr.phil. Friedrich] Hempelmann.

Winter semester 1933/34:

Spezielle Zoologie der Säugetiere (special zoology of mammals)—Prof.Dr. Grimpe
Allgemeine Geologie (general geology)—Prof. Dr. Kossmat
Spezielle vergleichende Histologie (special comparative histology)—Dr. Wetzel
Zoologisches Colloquium—Prof. Grimpe and Prof. Steche
Kleines geologisches Praktikum (basic course in geology)—[Prof.Dr.jur. et phil. Erich] Krenkel
Anleitung zu selbständigem wissenschaftlichen Arbeiten (instruction on independent scientific working)—Hempelmann.

Summer semester 1934:

Anleitung zu selbständigem wissenschaftlichen Arbeiten (instruction on independent scientific work)—Buchner [see box 3]
Erbkundliches Praktikum (course in genetics)—Grimpe
Übungen im Bestimmen von Blütenpflanzen (flowering plant identification course)—K. Wetzel
Pflanzengeographie von Deutschland (phytogeography of Germany)—K. Wetzel
Botanische Exkursionen (botanical excursions)—Bachmann and K. Wetzel
Mineralogie und Kristallographie I (mit Übungen) (course and lectures in mineralogy and crystallography)—[Prof.Dr.phil., Dr.phil.nat.h.c. Karl Hermann] Scheumann

Table 1. (*cont.*)

Mikroskopisch-optische Methoden (methods in optical microscopy)—Scheumann.

Winter semester 1934/35:

Tiergeographie (zoogeography)—Grimpe
Anleitung zu selbständigem wissenschaftlichen Arbeiten (instruction on independent scientific work)—Buchner
Physiologisches Praktikum für Naturwissenschaftler (physiology course for scientists)—Hempelmann
Zoologisches Colloquium—Buchner
Mineralogie und Kristallographie II (mineralogy and crystallography)—Scheumann
Mikroskopisch-optische Methoden II (methods in optical microscopy)—[Prof.Dr.phil. *Ernst* Karl Ferdinand] Kordes
Mikroskopische Übungen für Fortgeschrittene (advanced course in microscopy)—Bachmann
Geschichte der neueren Philosophie I (history of the recent philosophy)—[Prof.Dr.phil. Arnold] Gehlen
Spiel- und Wanderkurs (course in games and hiking)*
Praktikum Formenlehre (course in general morphology)—Hentschel [could not be identified with certainty, possibly Ernst H.].

Summer semester 1935:

Großes botanisches Praktikum (advanced botany course)—Ruhland
Spezielle Zoologie der Mollusken (special zoology of mollusks)—Grimpe
Geologische Kartierungsübungen (course in geological field mapping)—[Prof. Dr. phil. Carl Walter] Kockel
Spiel- und Wanderkurs (course in games and hiking)
Schwimmen für Nichtschwimmer (swimming for non-swimmers)—Prof. [Dr.phil. Hermann] Altrock
Anleitung zu selbständigem wissenschaftlichen Arbeiten (instruction on independent scientific work)—Buchner
Zoologisches Colloquium—Buchner
Russisch für Anfänger (Russian for beginners)—[Prof.Dr.h.c. Friedrich] Braun.

Winter semester 1935/36:

Vergleichende allgemeine Histologie (comparative general histology)—A. Wetzel.
Spezielle Zoologie der Fische, Amphibien und Reptilien (special zoology of fishes, amphibians, and reptiles)—Grimpe.

* The entry for this course was stamped with a mark saying: Course for candidates who aim at a teaching degree for secondary schools ("Staatsexamen"). Whether or not Willi Hennig indeed thought of acquiring this qualification cannot be decided. Most probably, he did not aim at becoming a schoolteacher, but regarded it clever to acquire the "Staatsexamen" before receiving his PhD (Doc. 203).

basic zoology course nevertheless. However, Schlee (1978) reports that Hennig was admitted to the advanced zoology course already in his second semester, which means that he had been dispensed with the two-semester basic zoology lecture (possibly this pertains to the lecture on general zoology by professor Meisenheimer, and Willi Hennig had to attend it only in his first semester).

The notification of scholarship by the Studienstiftung des Deutschen Volkes of 1932 is written in a completely sober, official style. In contrast, a circular of 10.07.1933 (Doc. 020), less than six months after the National Socialists had seized power on 30.01.1933, signals unanimous conformity with the Nazi ideology. Without any sign of regret the addressees are informed that from now on only Aryan students will receive grants. Moreover, it is emphasised that the participation of the awardees in military sports (Wehrsport) and Labour Service (Arbeitsdienst) should base on enthusiasm rather than obligations.

Willi Hennig had to deliver a report to the Studienstiftung on his first three semesters, i.e. at some time in autumn, 1933. He took this report seriously enough to write by hand at least two drafts, of which five pages the family preserved and kept (Doc. 021). Willi had written a first version which had been checked and annotated by another person (perhaps Fritz van Emden), and a second one, though without comments in another handwriting, but still with his own corrections. In this report drafts, Willi Hennig verbosely found excuses for the fact that he had not yet attended the—officially voluntary, in fact obligatory—Labour Service. He argued that he thought that he would serve his people most efficiently if he did what he could do best, and this being science. Other, more competent people (he did not mention names) were of the same opinion and had encouraged him to invest more time in developing good science skills than in non-scientific enterprises. Additionally, he had been at the Landesschule under an educative regime emphasising sportive activities to a high degree, so that he felt he was already physically well trained. However, he pointed out that he would not principally refuse to fulfil the new requirements and promised to do his Labour Service in the near future.

Finally, he completed the required service from 01.08.–01.10.1933 in a labour camp at Rodewisch (ca. 50 km South-West of Chemnitz, Saxony), and from 02.–15.10.1933 at Sayda-Kreuztanne (ca. 40 km South-East of Chemnitz), in the Labour Service groups 164 and 155, respectively (Doc. 022). The passport photograph of that time (Doc. 022) shows an elegiacly looking young man wearing a shirt with an open broad collar, who could rather be a member of the—at that time already forbidden—"Wandervogel" youth movement than of any of the National Socialist organisations (Fig. 20). Nevertheless, he joined the SA (National Socialist "Sturmabteilung") on November 5, 1933 (Doc. 173: Personal-Fragebogen). As an SA-man, he was assigned to the SA-Sturm 9 [in III. Sturmbann] of the SA-Standarte 107, which was based in Leipzig (Doc. 199). This membership does not show up in any other document, and since it is not even mentioned in Willi Hennig's military papers, it is certain that he did not hold a higher rank than the basic one ("SA-Mann").

While a student at the Leipzig University, he came regularly to the Dresden museum and visited Fritz van Emden and Wilhelm Meise. Together with the latter, he completed the two already mentioned papers on "flying" snakes, and in close co-ordination with Wilhelm Meise he wrote the two papers on the genus *Draco* (also "flying" reptiles), which were published in 1936. Certainly inspired by Fritz van Emden, he invested every spare hour in the taxonomic revision of the fly family Tylidae (nos. 2, 4, 8, 16, 18). Although some of these papers were published after his doctoral

Fig. 20. Portrait photo of Willi Hennig, from his Labour Service ID card, issued 23.10.1933. Courtesy of Gerd Hennig, Tübingen.

promotion, he must have done the empirical work in parallel to his university courses. This is shown by a letter of Fritz van Emden of 23 November, 1932, to Willi Hennig (Doc. 023), written as a reply to a letter of 17.11.1932 in which Hennig had asked for advice on a manuscript on Tylidae which he wanted to submit to a scientific journal. Fritz van Emden suggested to send it to *Stettiner entomologische Zeitung*, and mentioned as an alternative *Konowia*, an Austrian journal published and edited by the Viennese entomologist Fritz Wagner. Fritz van Emden suggested some technical and stylistic changes, and asked Hennig to consider publication of the immensely long paper in several parts. Indeed, the first section appeared in three parts in *Stettiner entomologische Zeitung* (no. 4) in 1934/35, and the second in five parts in *Konowia* (no. 8) in 1935/36.

Fritz van Emden had lauded Willi Hennig for this work, and confirmed that he rated it as belonging to the upper ten percent of all scientific manuscripts he had seen. Judging from Willi Hennig's semester report to the Studienstiftung, he knew very well that he was outstanding. Seemingly—and absolutely justified—he was proud of his achievements. Obviously, he applied for an extension of his grant, because there is the carbon copy of a letter of reference by professor Georg Grimpe to the Studienstiftung (Doc. 024). In this letter, Prof. Grimpe emphasised that he considered Willi Hennig's publications as being superb, and he wrote as a recommendation that he had invited Willi Hennig to give a one hour lecture on the superspecies- ("Rassenkreis") concept in the institute's regular colloquium, and that this lecture was excellent in every respect. Because this talk was such an outstanding achievement, he suggested to Willi Hennig to reorganise the snake collection of the zoological institute that contained about 300 species.

Two additional aspects are remarkable in this letter of reference. First, Prof. Grimpe did not spend even a single word on Hennig's political attitude, although exactly during the period in which he wrote the letter the Studienstiftung had fanatically turned to National Socialism. This is the more significant since Georg Grimpe signed on 11.11.1933 a public commitment to Adolf Hitler (Doc. 025). Thus, if he had thought that any mention of an affinity of Willi Hennig to the Nazi ideology could have enhanced Willi's chances to receive the extension of his grant, he certainly would have mentioned it. Further, in the closing paragraph of his letter Prof. Grimpe described Willi Hennig's character in some very pointed sentences: "As far as the purely human qualities of Mr. H. are concerned, I wish to point out that he gives a somewhat shy and unadept impression on first sight, but he gains ground as a person more and more, the longer and the more precisely one knows him" ("Was die rein menschlichen Qualitäten des Herrn H. betrifft, so möchte ich darauf hinweisen, dass er auf den ersten Blick hin einen etwas schüchternen und ungewandten Eindruck macht, daß er als Mensch aber mehr und mehr gewinnt, je länger und genauer man ihn kennt").

As usual in those times, Willi Hennig had to attend also certain non-scientific courses or lectures, especially in philosophy. It can be inferred from his record of study (Doc. 016, see Tab. 1), that he did little more than the minimum, possibly only that. The only entry explicitly labelled "philosophy" was a lecture on the history of younger philosophy by Prof.Dr. Arnold Gehlen in the winter semester of 1934/1935. Since the only other course in the faculty of philosophy listed in the record of study (Tab. 1) is "Russian for beginners", near the end of his studies, we can fairly conclude that Willi Hennig's professional training in philosophy could have been only basic, at best.

Already in his fourth semester, he received "instruction on independent scientific working", which is a cryptic circumscription of "accepted as a doctoral candidate by a professor". Almost certainly, Willi Hennig had

Fig. 79. Paul Buchner ca. 1926 (Courtesy of Zoologisches Institut und Museum der Ernst-Moritz-Arndt Universität, Greifswald).

planned to be supervised by professor Meisenheimer, who regrettably died in February, 1933. Most likely Friedrich Hempelmann took this task as a temporary replacement and filled the gap until Meisenheimer's successor, Paul Buchner, was appointed. Buchner took over to supervise Hennig's doctoral project, but did later not claim that he had initiated it, instead, he emphasised that either Meisenheimer or Grimpe was the actual scientific originator (Doc. 095). His own field of work was quite different from Willi Hennig's interests (see box 3). Yet, as it was at German universities

Box 3 Paul Buchner

Paul Ernst Christof Buchner was born on 12.04.1886 in Nürnberg, Germany (Fig. 79). His father was a physician and an amateur botanist who had a superb knowledge of local and Mediterranean flora (Anonymus 1956; Gersch 1979). Through him, Paul Buchner already became enthusiastic about nature as a boy. He entered the University of Würzburg in 1905 and studied biology with a focus on botany. However, under the influence of Theodor Boveri's, Johannes Sobotta's, and Philipp Stöhr's lectures he decided to orient towards zoology. He moved to München university in 1906, where he did his doctoral thesis on a cytological subject, inspired by Richard Goldschmidt but formally supervised by Richard Hertwig, for which he received his PhD in 1909 (Anonymus 1956).

Following his graduation, he worked for one year at the biological station of Naples in Italy. This period had a crucial impact on his further life. He developed a strong fondness for the landscape, flora, fauna, and people of that area. He later met his wife Miliana there, bought a house—the "Casa Buchner"—on the island of Ischia in 1929 (Stammer 1946), where he later lived. He returned to München, began to work on symbioses, and published a *Practical Course in Cytology* ("Praktikum der Zellenlehre") in 1915. Soon he focused on intracellular symbioses and published several books on this subject.

In 1912 he received his habilitation (ability to teach, the formal prerequisite to become appointed professor) at München University, was appointed "Extraordinarius" (associate professor) there in 1919 and full professor at the University of Greifswald in 1923. Here, he organised the overall renovation of the institute building from 1925 onwards (Kämpfe & Michalik 2011). In 1927, he became director of the Zoological Institute at the University of Breslau, but was called to Leipzig as the successor of Johannes Meisenheimer in 1934. As a professor at Leipzig, he was responsible for a lecture series on "General Zoology", from which a widely known book originated in 1938.

Even during his time at Breslau, the Buchner family regularly spent their holidays on the island of Ischia, and from 1943 they lived there permanently. Paul Buchner was detained on Ischia by the allies in 1944 and stayed there after his release. He had a private lab in his house and continued to study insect symbioses, especially those of scale insects (Gersch 1979). He died on Ischia on 19.10.1978.

in those times, even if a lower ranking scientist had the formal right to supervise a doctoral project, it was always advisable to choose the head of the institute as supervisor. Thus, a cytologist formally supervised Willi Hennig although he did a conventional morphological and taxonomical study on the copulatory apparatus of the Diptera Cyclorrhapha (a certain subdivision of the flies). At any rate, he must have developed a fairly good personal relationship with Paul Buchner who wrote him a letter in 1958 from the island of Ischia, assuring him that he understood his depressing feelings because of the life in Berlin at that time (Doc. 026).

Fritz van Emden's successor since 1934 at the Dresden Museum was Klaus Günther (see box 4). Willi Hennig must have come into contact

Box 4 Klaus Günther

Klaus Alfred Günther was born on 07.10.1907 in Berlin-Wilmersdorf as son of the Presiding Judge of the District Court Alfred Günther and his wife Elfriede, née Volprecht. He had a twin brother, Ulrich, and a younger Brother, Eberhard, on whose birth in 1911 his mother died. He attended the classical Gymnasium of Cottbus (Germany) until 1926 and studied biology, palaeontology, geography, chemistry and numismatics at the München University. From winter semester of 1927/1928 he was inscribed at the Friedrich-Wilhelms-Universität of Berlin, where he received his doctoral degree on 17.07.1931 for his dissertation on morphology and function of the mouthparts of the crustacean family Cymothoidae (Isopoda), supervised by Carl Zimmer. After a three years internship at the Berlin natural history museum, he was appointed head of the entomological department at the Dresden museum of "Tierkunde, Rassenkunde und Völkerkunde" (zoology, race science, and ethnology) in 1934, as successor of Fritz van Emden. He held this position until 31.07.1946. In 1935 he married Hildegard Kaufhold (who died on 06.01.1969). From 1942 until 1946 he was also in charge of the state collection of coins (Staatliches Münzkabinett). From 23.08.1944 to 09.05.1945 he served as artilleryman in the army. For three years he had no regular income but worked as a volunteer at the Berlin natural history museum. Here, the ornithologist Erwin Stresemann who managed to get him a position from 01.01.1948 as an assistant at the institute of genetics in the department of Hans Nachtsheim fostered him.

Jointly with Hans Nachtsheim, Klaus Günther moved to the newly founded Freie Universität of West Berlin on 01.04.1949. Here, he received his habilitation on 08.11.1950 and became "Extraordinarius" (associate professor) with the winter semester of 1955/1956. On 29.01.1960 he was appointed full professor of zoology and was head of the institute of zoology, jointly with Werner Ulrich, until he retired early in 1970. He married Waltraud Wolf in February, 1970, and died on 01.08.1975. He was buried in Berlin-Lichterfelde (Herter & Strübing 1975; Urich 1975) (Fig. 80).

Klaus Günther published 130 zoological papers, among them revisions of several hundred pages on Orthoptera, Mantodea, and Curculionidae, and a number on the functional morphology of the fish gnaw apparatus, plus a quite peculiar and artistic book—jointly with Kurt Deckert—on deep sea animals ("Wunderwelt der Tiefsee") in 1950. A French translation appeared in 1952, an English one in 1956, and a Dutch one in 1959. His most important contribution to the progress of zoology are certainly the two extensive review articles on systematics and evolution of animals in 1956 and 1962 in which he reviewed the entire phylogenetic and taxonomic literature of the period between 1939 and 1959. These reviews helped crucially in propagating the Synthetic Theory of Evolution and Willi Hennig's Phylogenetic Systematics in Germany. He was a close friend of Willi Hennig since their common time at the Dresden museum, and he supported him substantially in the argument with Ernst Mayr in 1974 over the relevance and the theoretical basis of a phylogenetic classification (Schmitt 1996). They met at least twice every year and exchanged numerous letters and smaller notes—Willi Hennig's family preserves a file containing 71 letters and 16 postcards from Günther to Hennig, and nearly as many carbon copies of letters from Hennig to Günther, all written between May, 1964, and January, 1975.

His most seminal own contribution to evolutionary ecology and theoretical biology is his concept of the Ecological Niche as a system of interaction between organisms and their "Umwelt", published in 1950 (Schmitt 1987a, b; Sudhaus 1996).

Additionally, Klaus Günther published twenty scientific papers on numismatics, on the chronology and meaning of the giant stone statues of the Easter Island, on Byzantine art, and on problems of the history of symbols, among them an acutely inspiring treatise on the unicorn in 1967 (Schmitt 2004).

with Günther very early or even immediately after his appointment. Hennig thanks in his thesis Klaus Günther for providing him with specimens of flies not occurring in Central Europe. As the dean of the philosophical faculty of the Leipzig university approved on 07.06.1935 (Doc. 027) Willi Hennig's application of May 21 that he could hand in his dissertation already at the beginning instead at the end of his eighth semester, he had most probably begun the practical work on these specimens in 1934.

The final, oral, examination in the course of Willi Hennig's graduation took place on Thursday, December 19, 1935. The examiners were Prof. Paul Buchner (zoology), Prof. Wilhelm Ruhland (botany), and Prof. Erich Krenkel (geology) (Doc. 028). The oral examination was rated "Sehr gut" (very good), the thesis "gut" (good). He received his degree on April 15, 1936. At this time, Willi Hennig had already published eight scientific papers with more than 900 printed pages. This was definitely unusual.

Fig. 80. Klaus Günther ca. 1965. Courtesy of Waltraut Günther (1985).

When exactly and under which circumstances Willi Hennig met Klaus Günther for the first time, cannot be said with certainty. But we can take it for granted that the two developed a close friendship very soon. Günther thanked Willi Hennig in a letter of 25.05.1936 (Doc. 029) for giving him a copy of his doctoral thesis and wrote it would grieve him that he could not see him on coming Whitsuntide (31.05./01.06.1936). Willi Hennig worked from 24.03.1936 on for some months as a volunteer at the Dresden museum (Doc. 030; Schlee 1978), so the two must have met regularly, and they have certainly also discussed scientific matters, as can be inferred from the letter mentioned above.

2.4. *Find a Job and Found a Family* (Fig. 20)

Whether or not Willi Hennig received a payment for his volunteer work is unknown. At any rate, he was in need for a professional position. He applied for a grant with the Deutsche Forschungsgemeinschaft (DFG) in 1936. Originally, he planned to study fossil Diptera included in amber at the amber museum of the Albertus-Universität at Königsberg (East Prussia) and at the zoology museum of the Berlin University. The first letter he wrote to the DFG is dated July 14, 1936. Hans Kummerlöwe, at that time director of the Dresden museum, and also Paul Buchner, and Walther Horn, director of the Deutsches Entomologisches Institut (DEI) in Berlin strongly supported his application. Walther Horn sent a letter of support to the DFG on November 12, 1936, in which he confirmed that Willi Hennig was diligent and sedulous ("fleissig und gewissenhaft"). At that time, Willi Hennig was in urgent need of financial aid. His parents had took up a loan from the bank in order to support him and had now to return the money, as Hennig explained in a pressing inquiry to the DFG, dated December 1, 1936 (Doc. 173).

Seemingly, Willi's desperate situation was known to other members of the staff at the Dresden museum, too. Obviously, there were colleagues who cared about the young post-doc (a term not yet known—in Germany—at that time). The keeper of Coleoptera, Prof.Dr. *Karl* Maria Heller advised him in a letter of Thursday, December 31, 1936, to ask Prof.Dr. Walther Horn, before January 8, 1937 for a position at the institute (Doc. 031). He recommended that Willi Hennig should emphasise that he was willing to become acquainted with any task in systematic entomology, even if it were to study insect larvae. It sounds as if Heller already knew that there existed such a possibility. Hennig wrote to Horn from his home address in Neusalza-Spremberg (Fig. 21), where his parents lived since

Fig. 21. The railway station building in Neusalza-Spremberg, where Willi Hennig's parents lived in the upper right part from the 1930s. Original, taken 20.07.21012.

around 1930, on 03.01.1937, a Sunday (Doc. 032). Already on the following Tuesday, Walther Horn offered Willi Hennig to work at the DEI (Doc. 033). The salary would, however, be extremely modest, at least at the beginning—110.- Reichsmark (RM), roughly equal to 390 € (Doc. 034)—but he offered to pay for the complete month of January if Willi Hennig would start working on Saturday, January 9. Interestingly, Walther Horn pointed out that Hennig would be listed in the staff rolls of the DEI as an awardee ("Stipendiat"), which had the consequence that Hennig would not have to pay social and health insurance fees. On 01.04.1937, Walther Horn confirmed (Doc. 035) that Dr. Hennig received a salary ("Gehalt") of RM 25.-, without giving a time period for which this amount was paid (probably one week). An entry in Willi Hennig's "Arbeitsbuch" (Doc. 036)—according to the law of 26.02.1935 all private and government employees had to have a permanent employment record—gives for the period of 08.01.1937 until 01.10.1938 at the DEI "additional professional training" ("Sonstige Fachausbildung"). This fact matters, because Willi Hennig stated in an official document (Doc. 037) that he had received a grant from the German Science Foundation (Deutsche Forschungsgemeinschaft—DFG) for this time.

According to the records of the DFG, they had given him a grant indeed, from April 1, 1937 to September 30, 1938. Interestingly, the project and his place of work had changed. It is obvious that Walther Horn, who esteemed the young entomologist, had guided Hennig to a slightly but crucially different path. In a handwritten letter of February 3, 1937, Willi Hennig explained to the DFG that he now intended to study in the first place larvae of beetles of economic importance in Germany, and only secondarily certain flies which could also be of practical relevance. This letter was regarded as a new application by the DFG and approved on April 5, 1937, retroactively from April 1 (Doc. 173). He received a grant of 2700.- RM for 18 months. On 01.10.1938, he became a regular assistant (Wissenschaftlicher Assistent) with an initial salary of 250.- RM (ca. 886 €) which should be raised to 300.- RM (ca. 1063 €) on 01.04.1939 (Doc. 038).

At the same time as Willi Hennig had inquired to Walther Horn for a chance to work at the DEI, he had also asked Carl Zimmer—at that time director of the Berlin natural history museum—for a job. Zimmer replied (Doc. 039) on January 13, 1937, that at the Berlin museum space was highly limited, so that not all scientists could have a personal office. Nevertheless, he signalled his principal willingness to help and offered to discuss the matter personally on the occasion of a meeting on the next Saturday.

Seemingly, Willi Hennig had used his parents' address only for the correspondence with Walther Horn. Carl Zimmer's letter was directed to an address in Berlin-Lichterfelde: Haydnstrasse 5 c/o Steger. At the time of his promotion, he was a lodger in Leipzig, in the house Beethovenstrasse 21, 1st floor. This was the address of the "Baurat" (government building officer, but the term means more a title than a profession) Paul *Ernst* Wehnert and his wife Bertha Frida née Sack. Their daughter Irma had studied biology and mathematics at the Leipzig University for five semesters, before she switched to history of art. Most likely Willi Hennig had met Irma Wehnert at the university and found a room as a lodger in her parents' house through her, but it cannot be excluded that he had found a room to rent and met her at the house of his landlord. According to an information given by Irma Hennig's niece Edith to Willi Hennig's son Gerd in August, 2012, her grandfather (Ernst Wehnert) normally never rented out rooms for students, and Irma and Willi knew each other well when Willi moved in (Doc. 040). Be this as it may, the two became a couple and married on May 13, 1939 in Berlin, although Irma's father was not entirely happy with his son-in-law (Doc. 167).

When his "Arbeitsbuch" (Doc. 036) was issued (07.10.1938), Willi Hennig's address was Moltkestr. 23 in Berlin-Lichterfelde. This area is a villa quarter with mansions and multi-family residences surrounded by generous gardens. The freshly married couple moved in a flat at Forststrasse 32 in Berlin-Steglitz, formally on the day of their marriage (Doc. 041). The flat was in a large block, located in an area more densely covered with buildings, but more central und urban than Lichterfelde. It remained their home until they fled the bombing near the end of WWII. Already before that time, Irma Hennig stayed at her parents' home in Leipzig from time to time for security reasons during pregnancy and for delivery, as seen from numerous letters from the secretary (Miss M. Blöink) and the interims director (Hans Sachtleben) of the DEI to Irma Hennig (Doc. 042).

The route from the flat at Forststrasse 32 to the Deutsches Entomologisches Institut (at that time in Gosslerstrasse 20) in Berlin-Dahlem was 2.7 km long and led through remarkably green areas, along private gardens, parks, and the Botanical Garden. The institute building (Fig. 22) lies in a remote street in the neighbourhood of exclusive mansions in the midst of alleys lined with beautiful trees. Thus, one can imagine that Willi Hennig enjoyed working at this institution, and he must have loved commuting to and from work, certainly walking (Doc. 172), since there was no comfortable bus line. If he went by bicycle—which we do not know, but is highly unlikely (Doc. 106, Doc. 172)—the bumpy cobblestone pavements could

Fig. 22. Gosslerstr. 20 in Berlin-Dahlem, institute building of the DEI from 1912 until 1943. Original, taken 18.05.2012.

have been the only encumbrance. According to Professor Wolfgang Hennig, his father commuted probably by underground, riding from Breitenbachplatz to Thielplatz (Doc. 175).

In 1933, a co-operation of the DEI and the Biologische Reichsanstalt was established (Doc. 104). This institution was located in a building in Königin-Luise-Straße 19 in Berlin-Dahlem. The two institutions jointly published the journals *Arbeiten über morphologische und taxonomische Entomologie* and *Arbeiten über physiologische und angewandte Entomologie aus Berlin-Dahlem*, in which Willi Hennig placed 23 papers until the end of WWII.

Even if this sort of life was what he had aspired to, he aimed at a more responsible—and better paid—position at a natural history museum. In March, 1939, he received a letter from Max Beier (Doc. 043), entomologist at the natural history museum of Vienna, in which Beier wrote "I think I have heard that you would not unwillingly come to our museum" and advised him to inquire decently at the director, Prof.Dr. Otto Pesta, for an available position. Willi Hennig did so (Doc. 044), and amusingly he mentioned in his letter that he was bound ("verpflichtet") to work on beetle larvae, thus clearly indicating his low motivation. Pesta's answer

(Doc. 045) was negative, however it seems that he really regretted that *currently* (emphasised by underlining in the letter) there was no chance to fill a position, because he had to wait until the officials in Berlin had decided upon the new staff appointment scheme (Austria had been annexed into Germany on 12.03.1938).

In 1937 and 1938, 20 publications authored by Willi Hennig were released from the press. Of course, quite a number had been practically prepared before his appointment at the DEI. On the other hand, most of the empirical work for the five papers published in 1939 must have been completed already in 1938. In his application letter (Doc. 032), Willi Hennig had followed the advice of Karl Heller and assiduously affirmed that he was willing to do any work on insects, even if it should be on larvae. Indeed, in 1938 he published a review on the larvae of the most important German Chrysomelinae (a subfamily of the leaf beetles). This remained the only paper on beetle larvae by Willi Hennig. Among the other 19 papers of these two years, there were five chapters of the famous "Fliegen der paläarktischen Region", edited by Erwin Lindner ("the Lindner"). From the letters to Walther Horn (Doc. 032) and to Otto Pesta (Doc. 044) we can clearly infer that Willi Hennig was well aware of the scientific importance of these chapters. It was indeed an extraordinary honour that he was chosen as the successor of Friedrich Hendel, who had originally taken over to contribute these chapters but had died in 1936. Among dipterists, Willi Hennig's name must have had a high reputation already so shortly after he had received his PhD.

Willi Hennig was called for an abridged military training from 06.02.–06.05.1939 (Doc. 037). He did this service at the Walter-Flex-baracks in Landsberg an der Warthe (now: Gorzów Wielkopolski, Poland) in the "6. E. / I.R. 50" (infantry regiment no. 50, Doc. 044).

Already in 1938, he had published a paper on flies fossilised in amber, and in 1939 a short paper followed on a flea fossil in amber. Here lay the roots of his intense study of amber fossils that should calender his scientific work thirty years later.

On 10th of July, 1939, Walther Horn died. Hans Sachtleben became the provisional head of the DEI, and was director from 1943 on (until 1962, Ebert et al. 1986). As can be judged from later correspondence—after WWII—between Sachtleben and Hennig, Sachtleben must have recognised Hennig's abilities, knowledge, and coming influence quite early (Fig. 23).

Fig. 23. Hans Sachtleben on an insitute's excursion ("Betriebsausflug") to the Woltersdorfer Mühle near Berlin in 1955. Courtesy of Gerd Hennig, Tübingen.

2.5. *An Entomologist at War* (Fig. 24)

On August 28, 1939, Willi Hennig was called up for the Army (Fig. 25). His unit was the 6th company of the infantry regiment no. 323 (Doc. 046), a subunit of the infantry division 218 (Doc. 047). This division was formed on August 26, 1939, and remained near Berlin (at "Wehrkreis III") even after the start of WWII (on September 1, 1939) before it marched behind the front to Thorn (now Toruń) in Poland (Doc. 048). Willi Hennig's army postal letters show that at least his company did not leave Germany before November 12, 1939. All 12 army postal letters plus one postcard written by Willi Hennig (Doc. 049) in 1939 and kept by the family are definitely sent from a place in Germany near Berlin, probably Zossen (since he mentioned several times a short leave that allowed him to see his wife, and once he described a return ride from home to the barracks by train). The name "Zossen" is written by hand on the back of a photograph (Fig. 24). Hennig's military unit was obviously not designed for serious military actions. In

Fig. 24. Willi Hennig as a recruit in 1939 at Zossen. Courtesy of SDEI.

Fig. 25. Willi Hennig as a soldier in September, 1939. Courtesy of Irma Hennig (1995).

his letter of 30.09.1939 to his wife he writes that they still were in the barracks and that at least for the next times to come this will remain the case. All his “new comrades” were older than 40 years.

This first phase of Willi Hennig's career as a soldier must have been rather a boring than an exciting or a dangerous period. He mentioned several times that there was little to do, and he asked Irma to send him books. On one hand, he complained—in a letter of 09.10.1939—about the guard service he had to do, but on the other hand he confessed that he enjoyed the four hours between each two-hour service because he could then write scientific papers. He worked in every spare hour on further chapters of Lindner's “Fliegen der Paläarktis”. He sent whatever he completed to his wife and asked her to forward it to Lindner. Likewise, he expected her to collect the galley proofs from the Deutsches Entomologisches Institut and send them to him.

On 22.10.1939, he asked Irma to have a look at a turnable stellar map which his former teacher Willy Matthes possessed and whom Irma was going to visit during her next stay at Leipzig. He did so because he wanted to have such a map himself, and he asked Irma to buy one and send it to him under his army postal service number. On 09.11., he confirmed receipt of this celestial chart, and on a photograph of that time he can be seen sitting on a bunk bed and studying position of stars by means of this map (Fig. 26).

In his letter of 27.10.1939, he inquired about a book he had seen on Irma's bookshelf—Wilhelm Wundt's “Allgemeine Geschichte der Philosophie” (General History of Philosophy)—and suggested that she could send it to him. In the same letter, he instructed her to call at the DEI and ask for sending him Lewis Wendell Hackett's book “Malaria in Europe—an Ecological Study”. From this correspondence we learn that Willi Hennig was interested in philosophy—still or already—in 1939, and that he regarded Malaria in Europe a topic worth being studied in these times.

In the last letter of 1939 (12.11.), he expressed his apprehension that after several false alarms during the past weeks they really had to march off in the near future. According to the “Lexikon der Wehrmacht” (Doc. 048), the Infantry Regiment no. 323 belonged to the 218th Infantry Division which was a reserve of the 4th Army that went to Poland behind the front in 1939 and remained there as an occupation troop until mid-1940.

Willi Hennig carried a small pocket calendar with him, in which he entered short notes, often only single words, and sometimes only stenographic signs (Doc. 050). In this calendar he jotted down where his company arrived, stayed, or moved, and from time to time he noted smaller

Fig. 26. Willi Hennig sitting on a bunk bed in German baracks in November, 1939, studying his stellar map. Courtesy of SDEI.

but remarkable events. On 04.01.1940 we find the locality "Puławy", the name of a city ca. 130 km southeast of Warszawa. Four days later, he noted "1 Wanze" (one bug), on 11th of January "−34 °C" (Fig. 27). They stayed in Puławy until 24.01., when they moved to Łuczynów, only few kilometres away. Even here, he must have had considerable spare time, because he noted on 27.01 that he completed reading Hans Grimm's "Volk ohne Raum" of 1926, a volume of nearly 1300 pages that was exploited by the Nazi propaganda to justify conquest of areas East of Germany (Doc. 051). On 31st of January 1940, he had received the galley proofs of the tables of the Ulidiidae-section of Lindner's work. From February 8 to 10, he had to do service in the office. On February 27, he had to do guard service at Bąkowice, 60 km southeast of Breslau (Wrocław), from where he made an excursion to Brzóza-Głowaczów (Fig. 28).

On 25.04., his unit returned from Kozienice, ca. 100 km southeast of Warszawa, to Germany and arrived after three days at Möckern, ca. 30 km East of Magdeburg. He was on leave until May 17 and stayed in Altengrabow near Möckern—a large military training ground—until he was

Fig. 27. Willi Hennig in Poland, guarding the unit's vehicles, 21.01.1940. Courtesy of Gerd Hennig, Tübingen.

Fig. 28. Willi Hennig as a soldier in Poland, Bakowice 27.–30.03.1940. Courtesy of Gerd Hennig, Tübingen.

transferred to West Germany on 1st of June, 1940. Irma could see him the two days before he left.

Hennig's unit went to Heilbronn, from there to Freudenstadt in the Black Forest, presumably by train. Then, he mentioned explicitly that they had to march at night several times until they reached Lahr-Schmieheim on June 10. The next day, he was promoted to private first class (Gefreiter). On Saturday, June 15, Hennig crossed the River Rhein near Weisweil, and obviously from then on "real" war was around him: combat, explosions, attacks, prisoners.

He must have foreseen this, because on the last pages, originally blank, of his pocket calendar, he wrote a message to his wife, which is a touching document of a sort that certainly thousands of men wrote in those days. Willi Hennig wrote a remarkably sober letter to his wife. He showed no signs of exuberant feelings, no explicit words of love or fear. Presumably he could, like so many men of his generation—and certainly also of younger generations—not express his emotions properly. It would be indiscrete to reproduce the entire message, but two passages shed a light so characteristic on Willi Hennig's personality that I dare to give them here. First, he found it "ominous" that precisely in the first letters Irma and he had ever exchanged they discussed on the question whether it is justified to sacrifice the most beloved for one's own country (as Irma wrote) or not (as Willi thought). Secondly, he asked Irma to take care that Sachtleben (then director of the DEI), should a necrology become necessary, mentions that he (Willi) regarded all of his previous work only as a preparation for a deeper ingression into the general problems of zoological systematics ("nur als Vorarbeiten für ein tieferes Eindringen in die allgemeinen Probleme der zoologischen Systematik").

The day before France surrendered on June 22, 1940, Hennig mentioned that booty and prisoners were uncountable ("Beute und Gefangene unzählbar"). He stayed in Laveline-devant-Bruyères until July 5, and marched from there in stages to Stiring-Wendel, close to the German border in the Saar area, where he stayed until July 12. This is the last entry for that year.

The 218th infantry division was put on leave from July, 1940 to April, 1941 (Doc. 048). Willi Hennig returned to the DEI and continued working on Diptera taxonomy, mostly the chapters of "the Lindner". He was formally released from military service on September 4, 1940, but was called up again on March 6, 1941. His unit was garrisoned in Denmark (Doc. 052), first in Vordingborg, from April 18 in Haderslev/Hadersleben (Doc. 049). Interestingly, numerous messages Willi sent to his wife by military mail

(Doc. 049) correspond exactly with Götz Aly's (2006: 101) report that German soldiers contributed to the welfare of their families by sending goods from the occupied countries. He offered nearly in each letter to send haberdashery, sweets, wool and other seemingly attractive items.

Irma Hennig was pregnant, and Willi's letters became warmer and more concerned the closer the day of her delivery came. He asked urgently for news on Irma's state of health, and he reported that he had dreamed of her and of the baby as if it were already their third child. Their son *Wolfgang* Hans Günter was born on 3rd of May in Leipzig (where Irma stayed at her parents' home), and according to Willi's letter no. 19 of May 3, 1941 he had received the news immediately through a telegram that was read out to him on telephone. On 9th of May, he expressed his concern about the growing danger of bomb attacks on Berlin and recommended Irma and Wolfgang should stay in Leipzig for the time being. On June 17, in his letter no. 41, he confessed that he was afraid the war could expand and that the quiet times he had in those days were the calm before the storm. Indeed, on June 22 the German army invaded the Soviet Union.

During all the time, Hennig continued working on manuscripts and proofs. Several times he thanked Irma for proofreading, mostly of chapters of "the Lindner". Quite understandably, he aimed at returning to science (i.e. entomology) even under the circumstances of the war. Thus, he inquired about a position with the Military Institute for Tropical Medicine (Tropenmedizinisches Institut der Militärärztlichen Akademie) in Berlin (Scharnhorststr. 35). Fritz Peus, head of the entomology department of that institute (Steffan 1976) and in September of 1941 on duty in Greece, recommended him to contact a chief staff surgeon by name of Erich Martini at the institute, because he knew that this officer sought for a trained entomologist (Doc. 053). In addition, he gave detailed advice on how to write the application letter. Since Fritz Peus was a taxonomic specialist of fleas and mosquitoes, it is highly probable that he knew of Hennig's qualifications already. Nevertheless, the first reply of Erich Martini was disappointing and written in a condescending tone (Doc. 054). Martini was director of the Institute of Tropical Medicine at Hamburg but was appointed docent at the Berlin institute by the Wehrmacht in 1940 (Doc. 055).

Hennig stayed in Denmark until the 218th division was transferred to Cholm (Kholm) in Russia as part of the 39th army, Heeresgruppe Nord, in January, 1942 (Doc. 056). It was most probably in that area where Willi Hennig was severely wounded by shrapnel on his right upper arm and shoulder on January 31, 1942—he nearly bled to death (Doc. 201). For that reason, he received the Batch of Casualties in Black (Verwundetenabzeichen in Schwarz) on 22.05., 1942 (Doc. 046), and on 15.08.1942 the so-called Ostmedaille (Doc. 057)—an honour awarded to all participants of the campaign against the Soviet Union in 1941/42 (Doc. 058). According to his notebook (Doc. 059), he was transported from the place where he was wounded to the military hospital of Allenstein (now Olsztyn) in East Prussia where he arrived on February 11 and stayed until February 19. He passed Berlin on February 20 and reached the hospital of Wanne-Eickel (ca. 10 km north of Bochum in the Ruhr-area). From there, he was transferred to the "Reservelazarett 101" in Berlin-Charlottenburg, where he was declared "ill from 06.03. until 02.07.1942" (Doc. 060). Already on February 16, the chief staff surgeon E. Martini had sent a considerably more polite letter to Willi Hennig at Allenstein inquiring if he could participate in a course in Medical Entomology running from March 4–27 1942 (Doc. 061). Shortly after, Hennig must have been promoted to Obergefreiter, as can be seen from the address of a letter by a captain at the Military Medical Academy by order of E. Martini, dated February 24 (Doc. 062). In that letter, this captain assured Hennig that the necessary steps had been taken to transfer him to Berlin.

According to Schlee (1978), Hennig was put on leave for half a year for work at the DEI and was thereafter required by the Military Medical Academy. The exact dates are unclear, but at least in May of 1942 he seemed to have been in military service. August Thienemann planned to edit a series on African freshwater fauna and asked Willi Hennig to treat some Diptera groups—they finally agreed on the Simuliidae (Doc. 063, Doc. 064, Doc. 065). Hennig completed a first version of the manuscript and sent it to Hermann Eidmann, director of the Institute of Forest Zoology at the University of Göttingen, who was responsible for the publication of the series. It was planned to appear in the "Schriftenreihe der Kolonialwissenschaftlichen Abteilung des Reichsforschungsrates—Fachgruppe Zoologie" but obviously never got to the press (Doc. 066, Doc. 067). Hennig was still affiliated with the DEI (Doc. 061) but received the further correspondence with Thienemann and Eidmann (Doc. 064, Doc. 065, Doc. 067) at his private address in Berlin-Steglitz (Forststr. 32).

Probably since July 2, 1942, Willi Hennig worked as an entomologist at the Institute for Tropical Medicine of the Military Medical Academy in Berlin (Doc. 174). On July 15, 1943, he wrote an army postal letter to Irma, one hour after he had learned of the birth of their second son, *Bernd* Dietrich (like Wolfgang, in Leipzig). He regretted that he was under military rule and could not take a leave at his convenience. According to the descriptions of his daily route to work, he had to commute between Berlin-Steglitz and Malchow in Berlin-Weissensee (now Berlin-Hohenschönhausen). There are no documents showing what he actually worked on—his publication record (Anonymus 1978) can only provide vague ideas. In 1942 and 1943, he published some applied papers on arthropods as transmitters of diseases, which might mean that it was his official job to study pest insects. However, also several basic taxonomic papers appeared, and one treatise of general importance, on the relations of larval and imaginal systematics (no. 63, 1943).

Letters to Irma from this time show that allied bombers flew attacks on Berlin, which became more and more a serious threat. Although Willi Hennig's style of language remained dry and unemotional, just the fact that he mentioned the bombing and that he described the damage they caused, demonstrates that the acts of war around him did not leave him mentally uninvolved. It is especially evident that he was seriously worried about plans of Irma to travel to Berlin by train.

While Irma Hennig stayed in Leipzig in September, 1943, Willi was given the rank of a "Sonderführer Z" and ordered to travel to Saloniki (Greece) for training in practical entomology ("Zur Ausbildung in praktischer Entomologie", Doc. 068). This military rank roughly corresponded to a platoon leader or Master Sergeant, it was normally given to staff or civil persons who had not passed the regular career but needed certain competences for special purposes (Doc. 069: 59). He left Berlin on September 30 and travelled via Vienna, Budapest, Belgrad to Saloniki/Thessaloniki, where he arrived on October 6 (Doc. 070). On his way, he visited the entomological museum collections of Vienna and Budapest and recorded meticulously the species of Simuliidae (Diptera) kept there. He took quarter in the former hospital of Náousa (Naoussa) near Saloniki and explored the area West, North, and Northeast of that town. He searched for sand flies (Phlebotominae), black flies (Simuliidae) and mosquitoes but collected also other Diptera which he found peculiar, e.g. some soldier flies (Stratiomyidae). He focussed especially on larvae and took some of them into his lab where he made rearing experiments. He made some extensive entries about records of *Mikiola fagi* (Hartig, 1839), a gall midge which he found on beech trees in the mountain nearby. Normally, he did not work on gall midges, but he had published a short paper in 1941 (no. 54) about parasitic Diptera, especially gall midges (Cecidomyiidae) and frit flies (Chloropidae), which might explain his sudden interest in that species in Greece.

According to his notes (Doc. 070), he had hardly any contact to other German entomologists during his stay in Greece, but he mentions some dinners and a Serbian wine feast, which he certainly had in company. He left Saloniki on November 16 and arrived in Berlin on 1st of December (Doc. 071).

Seemingly, one of his core fields of work were the sand flies (Phlebotominae). He must have prepared a comprehensive manuscript, as the Reichsforschungsrat (National Research Council) had granted 5000 RM on 29.03.1944 for its publication (Doc. 204). However, for some unknown reason the "Phlebotominae of the palaearctic region" never appeared.

Whether or not Irma and the two sons stayed in Berlin by then and if so, how long, is unclear. But from April, 1944, on Willi wrote from Berlin to Irma in Leipzig. In the course of the following weeks and months, he became increasingly concerned about the risks of travelling by train, due to the allied air raids. He also considered the pros and cons of Irma's possible return to Berlin and discussed several times the frequency, the strength, and the aftermaths of allied bombing of Berlin. Before July 2, 1944, his letters were regular ones, not sent by army mail. As we find no sign of enthusiasm or even confidence in a possible German victory in these letters, we can firmly assume that he was not in favour of the war at all. Since army postal letters could be checked at any time by military authorities and the Nazi secret police, the lack of these signs in later letters cannot mean much. He might have omitted this topic in later letters for safety reasons.

By end of June, he was sent to Upper Italy as a member of the staff of an Anti-Malaria unit ("Malaria-Lehrtrupp") within Heeresgruppe C. He left Berlin on June 30 (Doc. 070), made a stopover in Leipzig, where he met with Paul Schöps, publisher of books on applied zoology and animal breeding (Doc. 072), and met Fritz Peus on July 2 in Frankfurt am Main. He spent Tuesday morning, July 4, in the Diptera collection of the Stuttgart "Naturalienkabinett" (now State Museum of Natural History). It took him another 28 hours and a night in München to reach Verona. During his stay there overnight and also the following afternoon, there were air raid warnings, witnessing that he was in Italy not on a summer vacation. He travelled via Vienza—Padova—Adria to Mesola (near the coast of the Mediterranean Sea, between Venice and Ravenna), where his laboratory was, and arrived at his destination on Saturday, July 8. The following days, he explored the area along the coast, towards Venice, and to the South. He searched especially for sand fly larvae, but mentioned also other Diptera in his notes. Other than in Greece, he met up with colleagues several times who also served as military scientists in Southern Europe, e.g. Hans Krieg (at that time Lieutenant—Oberleutnant—and in civil life director of the Bavarian state collection of zoology) and Otto Niklas, specialist for sawflies and certain parasitoid Diptera. On August 11, he had to instruct an Italian medical officer on measures of insect control to be taken. Strikingly, he had to do this by help of an interpreter. That means that although he had learned some Spanish at school, he could—at that time—not speak Italian well enough to discuss a subject he was well acquainted with.

Formally, Willi Hennig was commander of a laboratory platoon ("Laborzug"), but in his notes is no mention of the number of staff he had. Instead, he wrote on July 5, when he was on his way: "Frontleitstelle Verona kennt Laborzug nicht" (front control centre does not know the laboratory platoon)—not just what one would expect of the famous German organisation. This could give the impression that the military operation was somewhat disordered at that time. However, this was seemingly not the case. Willi Hennig could routinely move around by bicycle and motorcycle, he confirmed receipt of equipment that he had ordered, and he could conduct experiments with different methods and chemicals to control mosquitoes. From time to time he had to get his bicycle repaired, and in his letters to Irma he regularly complained about the slowness of the army postal service. However, regarding this point, he was certainly unfair because he worried so much about Irma's and the boys' state of health that he found it "long" if a letter took more than ten days from Leipzig to his office in Upper Italy. Under the conditions of those times, it is rather astounding how efficient the postal service was—given the fact that to date a normal holiday greetings card needs at least one week to get from Italy's Adriatic coast to a smaller town in Germany. Admittedly, some letters got lost, and others arrived only after several weeks, but by and large the mail system ran as in times of peace. The main reason for Hennig's concerns was of a technical nature: He listened to the report on current air raids on the German broadcast and heard of bomber formations approaching or even bombing an area around Leipzig or the city itself, but could get news about his family only through postal letters which reached him always with considerable delay after the events they reported.

On September 21, 1944, he wrote to Irma that he had to work in Lignano, directly at the coast, and that he enjoyed taking a swim in the Adriatic Sea. He also reported in detail all the sights of allied airplanes he observed passing the area northbound, and he expressed his fears that these planes would going to bomb Leipzig. From time to time, he described that their office was attacked by machine gun fire from pursuit planes, and that bombs were directed towards targets nearby. Whenever there was alarm, he and his colleagues had to seek shelter in a slit trench. On September 9, he had sent a photograph to Irma, showing him as he sat at the entrance to that slit trench (Fig. 29). Seemingly, he was quite relaxed, smiled into the camera, as if there was no real threat at all. However, on October 15, he wrote that the civil hospital from which the slit trench was less than 50 m apart, was completely destroyed by bombs. His eldest son Wolfgang reports that he was was hit near the end of the war by a grazing shot and only saved by his steel helmet, which got dented on the side (Doc. 194). Possibly, this refers to a partisan attack that Willi Hennig later mentioned himself (Doc. 107).

In his notebook (Doc. 070), there are sketches of reports and recommendations regarding mosquito control. He had a tight schedule during the phase when he collected data and when he wrote the reports. But he must also have had some spare time since we find in the same notebook also hand-drawings of graphs representing either ideas on real phylogenetic relationships or expressions of his developing methodology, so to speak Hennig's "I think" (Fig. 30). He used several pages to lay out his considerations on the design of a registration card for museum specimens. For that purpose, he thought of creating an alphanumeric code, indicating the systematic position of a given species.

Fig. 29. Willi Hennig sitting at the entrance of a slit trench, Doná di Piave (Italy), 29.08.1944. Courtesy of Gerd Hennig, Tübingen.

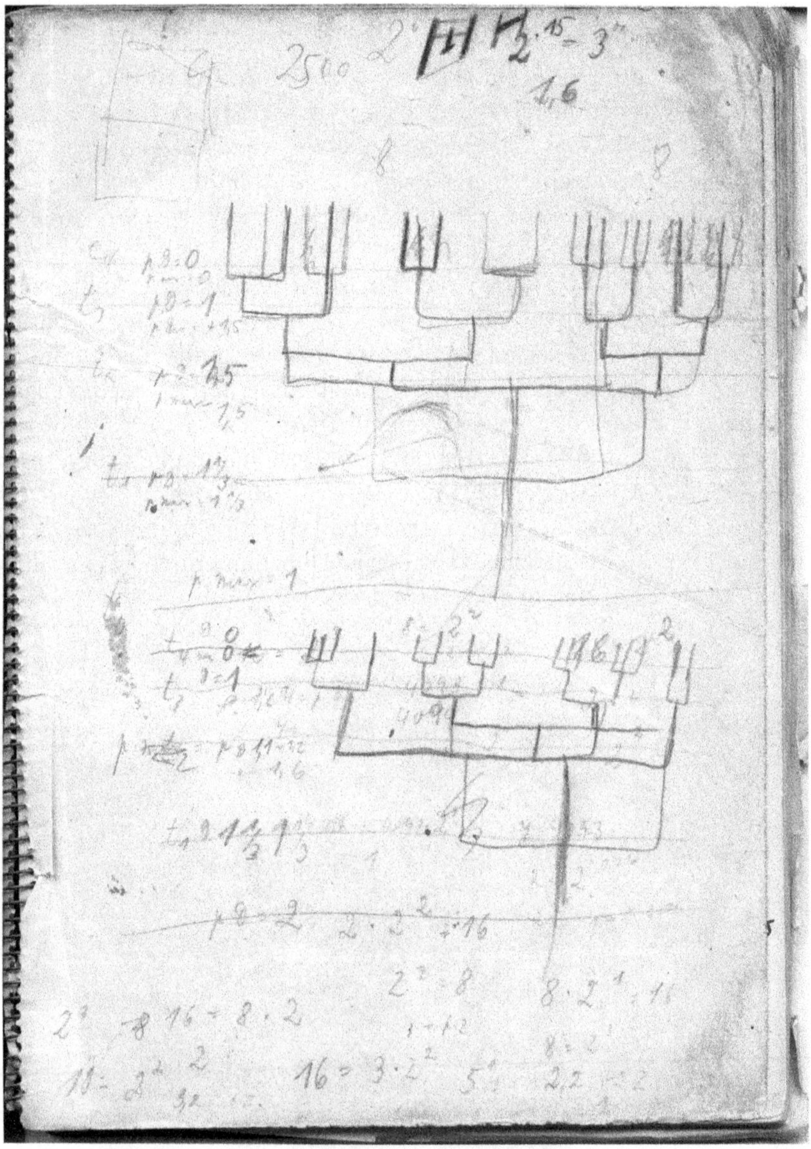

Fig. 30. Sketch of a cladogram in Willi Hennig's notebook, 1944. Courtesy of Gerd Hennig, Tübingen.

In a number of letters of November and December, 1944, he pondered over the chances of getting a leave over Christmas. Finally, he left Lignano on Friday, 15th of December, for Berlin. In all the letters written in 1944, Willi did not mention that Irma was pregnant again. Their third son *Gerd* Joachim was born on January 7, 1945 in Leipzig, while Willi was still in Berlin. The circumstances were dramatic, as Irma delivered her son during a bomb attack in her parents' house since she could not reach the bomb shelter in time (Doc. 201). Luckily, the house was not hit during this raid but only later. Hennig received a telegram on January 8, 14:45h, as he wrote in a letter to Irma on January 9. Remarkably, this telegram was postmarked at the post office in Leipzig on January 8, at noon. Even in the final phase of WWII, through ruins and damaged streets, a telegram was forwarded within a time not significantly longer than before the war. Amusingly, Willi confessed that when the telegram man handed over the message to him, his first idea was that it brought news about a suitcase he had sent from Italy to Berlin and which he still missed. Apparently, he was not up to date on the predicted time of his wife's delivery. Although he wrote this letter, as usual, in his decent and jejune style of language, it is clear that he felt relieved and that he was proud and happy, as he went immediately to neighbours to tell them the family news.

On his way back to Italy, he visited Irma and his boys in Leipzig on January 18, and arrived at Udine the next day. In some of the letters before and after that date, he had mentioned "my affair" without giving a clear account. A letter from his brother Rudolf of January 14 (Doc. 073) provides the solution: There had been plans to promote Willi Hennig, at that time Sergeant (Doc. 074), to officer. On March 6, 1945, he received the military cross of merit, 2nd class with swords. This decoration was destined for merits in the homeland or behind the front (Klietmann 1996: 37). The certificate is dated January 31 and signed by Albert Kesselring, commander of Heeresgruppe C. It is unclear whether a special event was the reason for this award (perhaps surviving the partisan attack in Lignano, Doc. 107) or if the bestowal was part of a routine preliminary to advancement promotion to officer. At any rate, there is no evidence that Hennig finally was promoted officer.

His letters to Irma written during the last four months of WWII reflect the growing danger for the population of German cities. Every time he wrote, he was worried about the well-being of Irma and their sons, and more and more often he mentioned his concerns about their "good Berlin" ("unser gutes Berlin"), and from February on also his anxiety about the fate of his parents. He could trace the progress of the Red Army towards

Saxony through the German broadcasting. As the Soviet troops had crossed the borders of the Reich in Silesia by end of January and approached Saxony, Willi was in growing fear for his parents (and scientific stuff he had stored in their house in Neusalza-Spremberg). Nevertheless, even in the last letter preserved (no. 18, dated 26.03.1945), he described routine work and announced that he had to pick up two new laboratory technicians at San Donà di Piave.

When the Heeresgruppe C surrendered on May 2, 1945, Willi Hennig was taken Prisoner of War by the British troops and brought on May 10 to a military hospital at Abano Terme, where he was deployed in a British anti-malaria unit (Schlee 1978) and obviously co-operated with the British (Doc. 075). Malaria was indeed a serious problem for the Allied troops (as it was for the Germans), as proven by the fact that the Allies had spent more than 470.000 £ for malaria control between January and September 1944 only in the regions south of the province of Grossoto (roughly south of 43° N, Doc. 076). In a report of April, 1945, it is stated "This disease, which has decimated armies in previous campaigns, was the greatest cause of sickness in CMF [Central Mediterranean Forces] last year" (Doc. 077).

According to short personal notes of that time in a pocket calendar (Doc. 168), Hennig was allowed to leave the institution. However, at least in the beginning a British officer always accompanied him (Doc. 078). From July 2 to August 14, 1945, he lay in the British hospital 651 (Doc. 174) at Abano Terme, as he suffered from jaundice (Doc. 078). I could not find out what exactly Willi Hennig's job was at Abano Terme. At any rate, he must have had some spare time or a liberal working schedule, as he was able to write down the manuscript of the book that later made him the founder of a Scientific Revolution (*Grundzüge einer Theorie der phylogenetischen Systematik*). He wrote in an Italian hard-bound notebook of 170 pages of squared paper of 21 × 31 cm (Doc. 079). Some additional pages were glued in later. He must have had a complete elaborated plan of the book in mind when he started writing, as can be seen from the handwritten table of content, the first page of the manuscript (Fig. 31). He wrote also the introduction nearly without corrections (Fig. 32), whereas on later pages he crossed out parts and added others (Fig. 33).

Willi Hennig must have been transferred to Plön (in North Germany) between September 30 and October 16, 1945. He had still received issue no. 114 of the Lager-Zeitung of Sunday, September 30 in Italy, and was discharged on October 16 from Plön. From Plön he proceeded via Göttingen to Dransfeld near Göttingen, where he received food ration cards

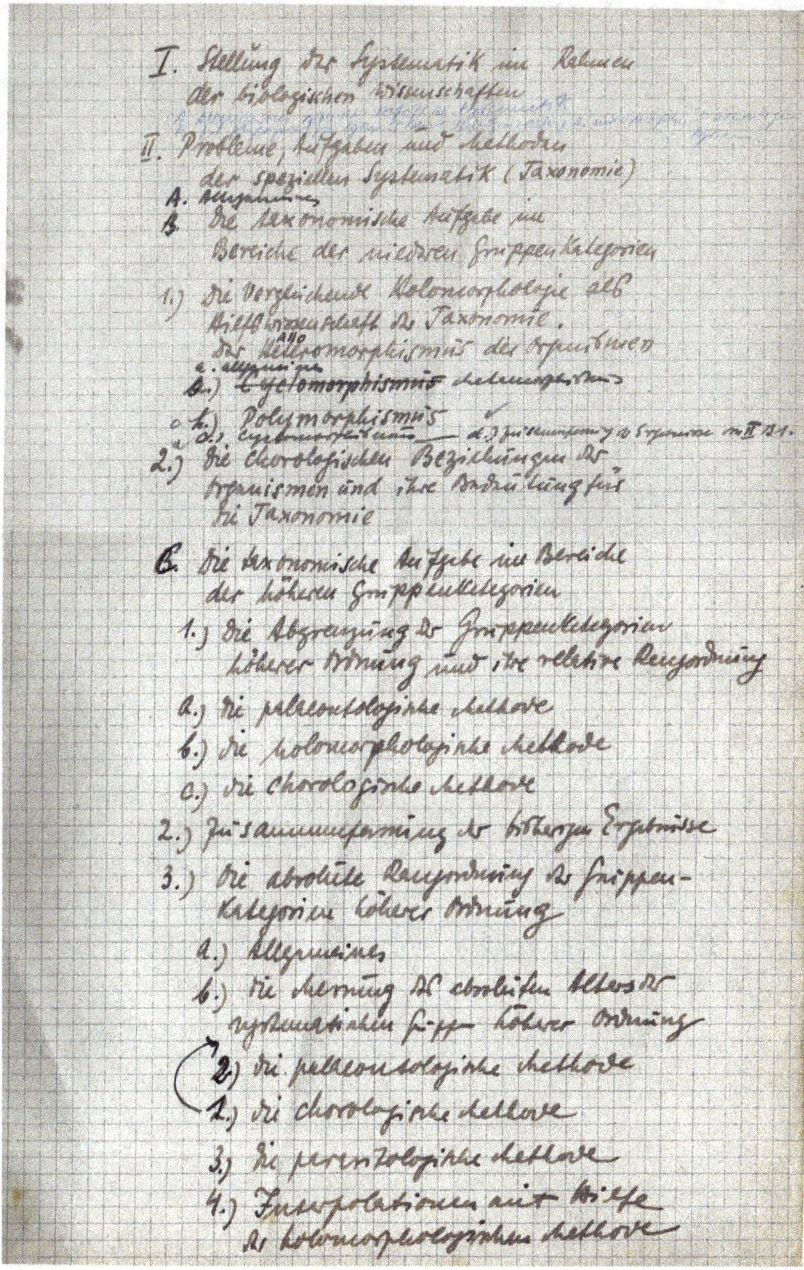
I. Stellung der Systematik im Rahmen der biologischen Wissenschaften

II. Probleme, Aufgaben und Methoden der speziellen Systematik (Taxonomie)

A.

B. Die taxonomische Aufgabe im Bereiche der niederen Gruppenkategorien

1.) Die vergleichende Holomorphologie als Hilfswissenschaft der Taxonomie.

b.) Polymorphismus

2.) Die chorologischen Beziehungen der Organismen und ihre Bedeutung für die Taxonomie

C. Die taxonomische Aufgabe im Bereiche der höheren Gruppenkategorien

1.) Die Abgrenzung der Gruppenkategorien höherer Ordnung und ihre relative Rangordnung

a.) die paläontologische Methode

b.) die holomorphologische Methode

c.) die chorologische Methode

2.) Zusammenfassung der bisherigen Ergebnisse

3.) Die absolute Rangordnung der Gruppenkategorien höherer Ordnung

a.) Allgemeines

b.) Die Bestimmung des absoluten Alters der systematischen Gruppen höherer Ordnung

2.) die paläontologische Methode

1.) die chorologische Methode

3.) die parasitologische Methode

4.) Interpolationen mit Hilfe der holomorphologischen Methode

Fig. 31. First page of the table of content in the manuscript of the *Grundzüge* (composed of two scans, as the original is larger than A4). Courtesy of Wolfgang Hennig, Kranenburg.

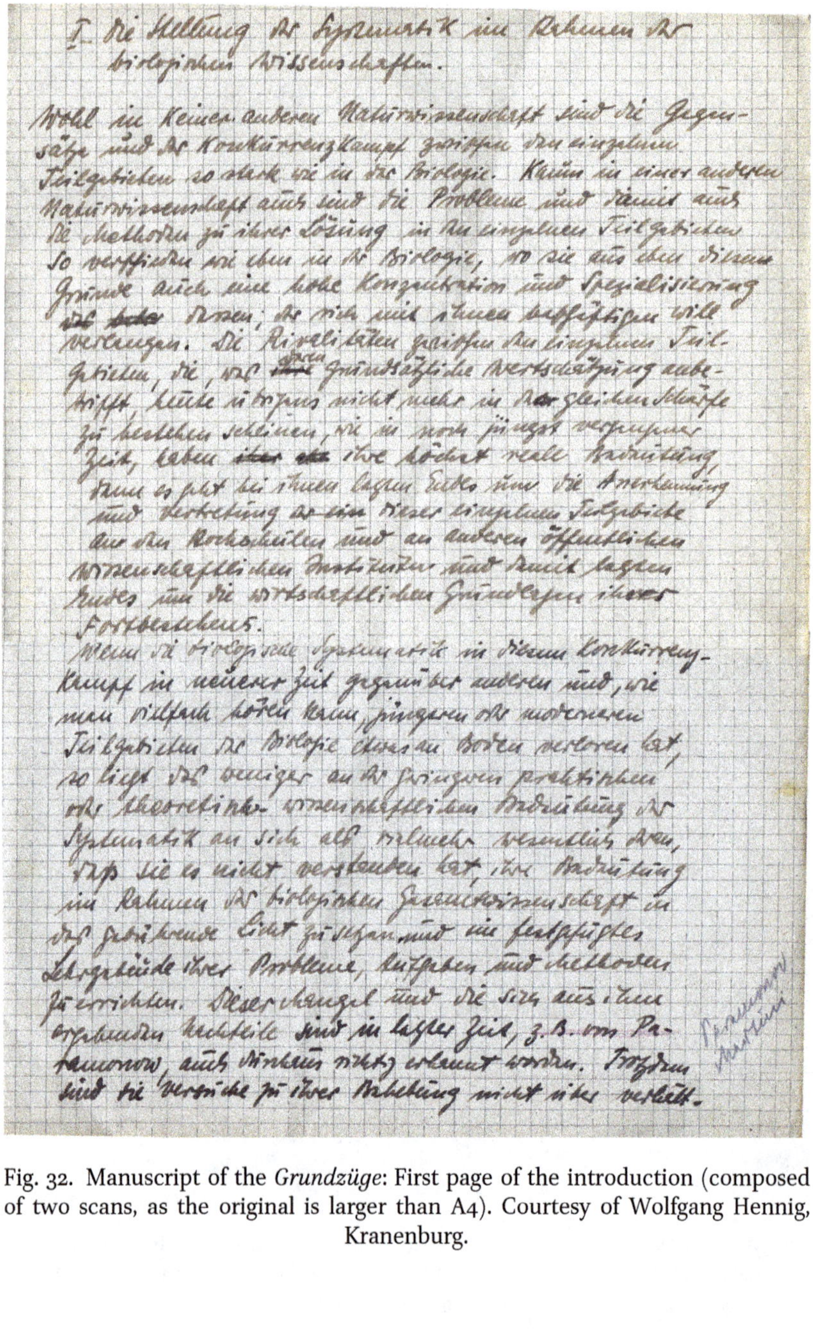

Fig. 32. Manuscript of the *Grundzüge*: First page of the introduction (composed of two scans, as the original is larger than A4). Courtesy of Wolfgang Hennig, Kranenburg.

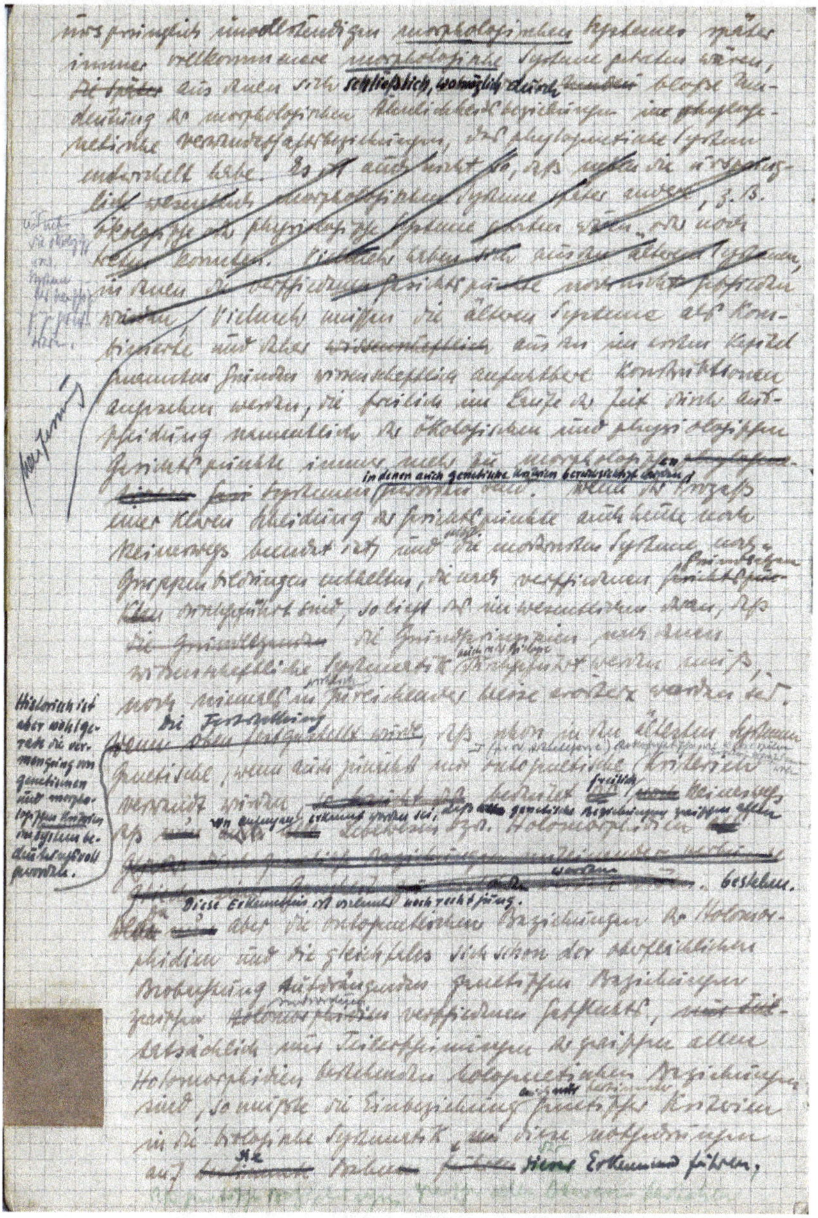

Fig. 33. Manuscript of the *Grundzüge*: Page corresponding to p. 18 of the book (composed of two scans, as the original is larger than A4). Courtesy of Wolfgang Hennig, Kranenburg.

for the time until November 27 (Doc. 080). According to Schlee (1978), Hennig took this route and crossed illegally the border to the Soviet zone by end of November because he feared to be taken POW again by the Soviets. Nevertheless, he wanted to reach his family in Leipzig and finally succeeded. When he arrived, he gave a considerably haggard appearance (Fig. 34). He could move in with his parents-in-law whose flat had been severely damaged but could be used (Doc. 167).

At the University of Leipzig, the position of Willi Hennig's doctoral supervisor, Paul Buchner, was vacant. Buchner had not returned from the island of Ischia since 1944 (see box 3), and his replacement Friedrich Hempelmann had been taken by the retrenching US Army to Weilburg/Lahn (Hessia) in 1945 (Doc. 016). Hennig was appointed stand-in for Paul Buchner (or Friedrich Hempelmann) on December 1 (Doc. 037) and gave courses until end of summer semester, 1947. The courses were: Allgemeine Biologie (General Biology) and Zoologisches Praktikum (Practical Course in Zoology) every semester, Spezielle Zoologie der Insekten mit beson-

Fig. 34. Willi Hennig in the photograph of his provisional passport, issued 01.12.1945. Courtesy of Gerd Hennig, Tübingen.

derer Berücksichtigung der medizinisch und landwirtschaftlich wichtigen Formen (Special Zoology of Insects with special emphasis on forms of medicinal and agricultural importance) and Zoologisches Praktikum für Anfänger (Basic Course in Zoology) in winter 1945/46, Zoologische Bestimmungsübungen (Zoological Identification Course) and Zoologische Exkursionen in summer, 1946, and a practical course on vertebrates in winter 1946/47 (Doc. 081). An amusing side note may be that the rector of the Leipzig University in those times was Hans-Georg Gadamer, a philosopher whose "Hermeneutics" comprised the so-called hermeneutic circle, an epistemological concept close to the "principle of reciprocal illumination" that Hennig exploited in his *Grundzüge*. Indeed, when I visited Hennig's widow Irma in 1995, I saw on his bookshelf a copy of Gadamer's *Wahrheit und Methode* (the first edition of which appeared, however, only in 1960).

Although the job at the Leipzig University would have guaranteed him a sufficient income to feed his family, he longed for his original position at the DEI. On February 26, 1946, Otto Schlumberger, then president of the Biologische Zentralanstalt für Land- und Forstwirtschaft wrote to Willi Hennig, obviously replying to an inquiry from Hennig (Doc. 082), that Hennig's former position as an assistant at the DEI would be kept open for the time being. Indeed, in March of 1947 Willi Hennig could inform the rector of the University of Leipzig that he would leave for Berlin (Doc. 083). From April 1, 1947, he could return to Berlin and continue his desired work at the DEI (Doc. 037). The family moved into their old flat in Forststrasse 32, Berlin Steglitz (Doc. 084) (Fig. 35).

2.6. *Willi Hennig's Time at the Deutsches Entomologisches Institut (DEI) after WWII* (Fig. 36)

The DEI had been removed from Berlin to Blücherhof manor in Mecklenburg (Fig. 37), ca. 170 km North-West of Berlin, in August 1943 (Ebert et al. 1986) in order to save collections and the library from the bombs. This manor has been bought in 1904 and completely reconstructed by Alexander Koenig, the founder of the Zoologisches Forschungsinstitut und Museum (now: Zoologisches Forschungsmuseum) Alexander Koenig (ZFMK) in Bonn (Hutterer 1998). After Koenig's death in 1940, his widow Margarethe Koenig intended to sell Blücherhof manor but did not receive a financially attractive offer. She died on May 14, 1943, and the executor of her and Alexander's will rented Blücherhof out to the Kaiser-Wilhelm-Gesellschaft, the

Fig. 35. Irma and Willi Hennig in their flat in Berlin-Steglitz, 1950s, taken by Wolfgang Hennig. Courtesy of Wolfgang Hennig, Kranenburg.

Fig. 36. Deutsches Entomologisches Institut, Berlin-Friedrichshagen, seen from the South, 1960. Courtesy of Gerd Hennig, Tübingen.

Fig. 37. Blücherhof manor, where collections and library of the DEI were housed from 1943 until 1950. Original, taken 20.10.2010.

head institution of the DEI at that time and also of the ZFMK, on June 28, 1943. Collections and library of the DEI moved in on August 4, 1943, the administrative headquarters of the Kaiser-Wilhelm Gesellschaft followed in April, 1944. Shortly after, the Gauleiter of Mecklenburg ordered to confiscate the manor for the Kaiser-Wilhelm-Gesellschaft. The Soviets expropriated the whole estate but let the DEI use the mansion (Hutterer 1998). After the end of WWII, the Kaiser-Wilhelm-Gesellschaft could no longer fund the DEI. Hans Sachtleben tried hard to manage that the DEI could survive, and finally the Biologische Zentralanstalt (the former Reichsanstalt) für Land- und Forstwirtschaft took it over on January 1, 1946 (Ebert et al. 1986). Probably, Willi Hennig had a provisional office in their institute building (Berlin-Dahlem, Königin-Luise-Str. 19, Fig. 38), but at least from time to time he had to work on Blücherhof manor. According to the memories of Hennig's eldest son Wolfgang, he had also a working place at the Natural History Museum of Berlin, where he used to work on Saturdays (Doc. 175).

Hennig could not commute from Berlin to Blücherhof on a daily basis but had to stay in the countryside for longer periods. From Blücherhof, he wrote letters to Irma of which two have survived (Doc. 085, Doc. 086). He praised the quietness of Blücherhof: no traffic, no visitors, no need to

Fig. 38. The former "Biologische Reichsanstalt", now "Julius-Kühn-Institut", Berlin-Dahlem, Königin-Luise-Str. 19. Original, taken 31.10.2012.

commute day by day. He must have lodged on the manor since there is no other possibility within a radius of 3 km, and he enjoyed that everything he needed for his work was conveniently at hand. However, he was concerned about the future fate of the DEI, and he worried about the food supply for Irma and the kids. He seriously pondered over transporting a sack of potatoes from Blücherhof to Berlin. These letters prove that he could travel from (West-)Berlin to Blücherhof although the Soviets had blocked Berlin's three West-Sectors right during that period of time (June 24, 1948 to May 12, 1949). This does not apply exclusively to Willi Hennig or other privileged persons, rather all citizens of West Berlin could freely enter the GDR and return to West Berlin during the blockade, in contrast to what one might expect on the basis of the partly dramatic descriptions of the conditions of everyday life in those times (Doc. 203).

The letter written on July 29, 1948, (Doc. 086) is a rare document of Willi Hennig's humour. Although he definitely had a good sense of humour and liked jokes and witty conversation (Schmitt 2001), he seldom expressed this in written form. Here, he jested that he had heard from Berlin to Blücherhof the boys bickering and finally confessed that he could not see if Vera (a niece of Irma) had come or not.

According to Schlee (1978), Hennig refined the manuscript of the *Grundzüge* when he was in Berlin, at candle light in a cold room. However, in the paper on *Probleme der biologischen Systematik*, published December, 1947, he explained in a footnote that publication of the *Grundzüge* was delayed due to paper shortage. Consequently, its manuscript must have been completed before that date. Possibly, the manuscript was submitted but Hennig continued working on it anyways. Finally, he received the first galley proofs in June, 1950 (Doc. 087).

In 1948, the first part, comprising 185 printed pages, of the three volumes on the *Larvenformen der Dipteren* (forms of dipteran larvae) appeared and became not only a highly estimated and sought-after contribution to the morphology and biology of Diptera, but—right for this reason—formed a crucial part of the stocks of the DEI library that could be offered for literature exchange. Publication of the previous journals published by the DEI which were the basis for the growth of the library through literature exchange—the *Arbeiten über morphologische und taxonomische Entomologie aus Berlin-Dahlem*, the *Arbeiten über physiologische und angewandte Entomologie aus Berlin-Dahlem*, and the *Entomologische Beihefte*—had to be ceased by end of 1944 (the former two) or even 1943 (the last). Therefore, Hennig's *Larvenformen* was soon a highly welcomed addition to the leftovers of the stock of Horn & Schenkling's *Index Litteraturae Entomologicae* (Ebert et al. 1986). Since financial means were extremely restricted in those early years after WWII, exchange of publications was the only possibility for the library to keep up with the increase of scientific literature.

After three years of working under the provisional conditions at Blücherhof, the laboratories, offices, and library of the DEI moved into a relatively ample accommodation in Berlin-Friedrichshagen, Waldowstr. 1 (since 1958: Josef-Nawrocki-Str. 10, now an elegant restaurant) between February 6 and May 23, 1950 (Ebert et al. 1986). The building is situated at the shore of Lake Müggelsee and provides a picturesque ambiance for scientific work (Fig. 36). Willi Hennig was in the first place responsible for the transfer and engaged himself physically. His son Wolfgang recalls that he accompanied his father riding on the lorry that carried books, collections, and equipment from Blücherhof to Friedrichshagen (Doc. 175).

On October 7, 1947, the family had moved into a flat in Opitzstr. 3 in Berlin-Steglitz (Doc. 088). The new flat was in the same block as the old one (Forststr. 32)—the Opitzstrasse runs parallel to the Forststrasse. The new flat offered more space—five instead of three rooms (Doc. 172)—and was nicer than the old one (personal communication by Gerd Hennig 2012), and perhaps it was more expensive so that the Hennigs could only afford for it after Willi Hennig's salary was raised. On November 1, 1949,

Willi Hennig advanced to head of the department of Systematic Entomology and vice-director of the DEI (Doc. 089).

In 1950, the *Grundzüge* were finally released from the press (Fig. 58). The DEI had subsidised its production and kept nearly the complete print run for literature exchange. The publishing house—Deutscher Zentralverlag (Berlin)—was not a distinguished publisher of scientific works and did—under the provisions of a socialist economy—not especially advertise this book. Consequently, it took years until a significant number of colleagues could read it, in the Western as well as in the Eastern world. In addition, Willi Hennig's style of language was a severe hindrance even for German-speaking readers. Obviously, he intended to express his thoughts as exactly as possible and wanted to leave no room for undesirable interpretations. Earlier papers of his as well as private letters show that he was not at all a brilliant writer. Thus, he ended up with long-winded sentences, used inconvenient terms, and dealt with single side-aspects in enormous detail. Another reason for the slow reception of Hennig's method was certainly that he was an entomologist working at an explicitly entomological institute (Schmitt 1996), and that the latter was nearly exclusively responsible for the distribution of the book. Since the DEI exchanged literature preferably with other entomological institutions, it took even longer until the *Grundzüge* found their way into fields outside entomology.

Although published later than the two small papers (1947, 1949) in which he outlined some basic terms and concepts of his method, the *Grundzüge* mark the starting point of a fundamentally new paradigm in systematics. I treat Hennig's historical contribution to the formation of what is now called "cladistics" in chapter 5. To summarise, he discussed in length the formal and philosophical background of systematics and classification—theory of sets, the "principle of reciprocal illumination", transformation theory and others. He showed that a fuzzy understanding of "relationship" prevents scientifically consistent hypotheses on natural order and therefore emphasised that the only rational solution is to restrict the concept of "relationship" to "genealogical relationship" which he called "phylogenetic relationship". The main part of the book contains numerous attempts to assess phylogenetic relationships by tracing character evolution, i.e. to find indications for polarising two or more states of a character in question. The backbone of his method is the statement that homology per se is not a proper indicator of phylogenetic relationship but only possession of shared transformed character states which he called "apomorphy". Fundamental for all following phylogenetic studies—of Hennig and others—was his concept of "monophyly". He did not coin this term, but he insisted that it is only useful in phylogenetics if it is defined as "comprising all descendants of a single stem species". The emphasis lies on "all" because since Haeckel's *Generelle Morphologie der Organismen* (1866) the term "monophyletic" was simply understood as 'descending from one stem' without specifying whether or not *all* descendants had to be included. Hennig's central claim was that a strictly phylogenetic system must contain monophyletic taxa exclusively (and possibly single species, but he did not explicitly mention this) (Fig. 39).

In 1951, he received the right to teach (venia legendi) jointly with the 'habilitation' (the confirmation of teaching ability) from the Brandenburgische Landeshochschule at Potsdam, on August 1 (Doc. 090). On October 10, the same year, he was appointed "professor" at this institution, which was rather a college than a university. He gave a four-hours-lecture on Special Zoology of the vertebrates, another of five hours (per week) on Systematic Zoology of invertebrates, a seminar on Systematic Zoology (one

Fig. 39. Willi Hennig in 1950, professional photograph. Courtesy of Irma Hennig (1995).

hour), and an identification course of three hours (Doc. 088). Seemingly, he had hoped that the DEI and the Brandenburgische Landeshochschule would develop a closer co-operation in the course of time, but this did not come true. So, he informed the rector at Potsdam by end of 1953 that he wanted to cease his teaching activities there. The reasons he gave were that agreements were not kept and that due to complete absence of any concessions he had to give his four- and five-hours lectures on one day each. This means either he was not allowed to leave the DEI for teaching on more than one day per week, or to give the lectures in Berlin instead of Potsdam (Doc. 091). Nevertheless, he kept the title of a professor.

The 1950s were an extraordinarily productive time for Willi Hennig. It seems as if he had eagerly waited for the chance finally to write and publish what he had in mind and could not get printed in the 1940s. He penned the remaining two volumes on the larval forms of Diptera, twelve papers on taxonomy and nomenclature of Diptera, three chapters of books on applied entomology, three papers on the phylogeny of insects, and—especially important—the *Taschenbuch der Zoologie* in two volumes. In the three phylogenetic papers (nos. 80, 89, 95) he exemplified the method of phylogenetic systematics. It was not just that he gave an extensive case study and by doing so made it easier for colleagues with a focus on practical science to understand his approach, but he also refined his methodology, introduced new terms, and described his method in a clearer style than in the *Grundzüge*.

The latter is especially true for the *Taschenbuch der Zoologie*, which comprises a stringent treatise of the protists and invertebrates. The two pocket books offer a concise overview on characters that could be regarded synapomorph for the taxa treated. Thus, the characters presented could form the basis for a complete phylogenetic tree of the animals. However, Hennig did not present any cladogram, and the book was designed and advertised as a textbook for undergraduates. Also, he did not provide until the fourth—posthumous—edition a general introduction in which he explained his methodical approach. Consequently, the scientific establishment did not take it as a serious alternative to common classifications, and the factual readers—the students—did not realise that they held a scientific innovation in their hands.

Willi Hennig's dear friend Klaus Günther praised the new phylogenetic method in a widely perceived review article in 1956, and again in 1962, and crucially helped making Phylogenetic Systematics known to an audience outside entomology (Schmitt 1996). Hennig's rising scientific reputation is reflected in several distinguished awards: He received the "Fabricius Medaille" (Fig. 55) for the year 1953 of the German Entomological Society

(Doc. 092) in 1954, was nominated Corresponding Member of the Finnish Entomological Society in 1955 (Doc. 093), and appointed regular member of the German Academy of Sciences (from 1972: Akademie der Wissenschaften der DDR) the same year (Doc. 094).

Since Willi Hennig lived in West Berlin while his office was in East Berlin, he could much easier attend congresses in Western countries than his East German colleagues. For example, he was invited to participate already in the 9th International Congress of Entomology, Amsterdam (The Netherlands), August 17–24; in 1951, at least he filled an official Dutch questionnaire (Doc. 096). However, most probaby he did not travel, as he kept the originally signed form. In 1958, he could attend the 4th International Congress of Zoology held in London (UK), 16–23 July, and see friends and colleagues on this occasion, among them Fritz van Emden (Doc. 097, Doc. 098). He could also visit museum collections in Western countries, but these visits had to be approved by the Academy for Agricultural Sciences of the GDR (e.g. Doc. 103: 25). However, travelling was not easy in those times, especially for the citizens of West Berlin. In order to reach the college in Potsdam, Willi Hennig had to pass the border to East Germany, and for this purpose he needed a permit which he could only get after producing his service certificate and his "Arbeitsbuch" (employment record), and he could cross the border only at one of three checkpoints (Doc. 099). The "Single Round-Trip Interzonal Pass" that allowed him to travel from Berlin to München and back in 1953 is marked by 12 stamps and bears equally many signatures (Doc. 100). In contrast, citizens of the Eastern part of Germany and of East Berlin could hardly travel to Western countries or to the Western zones of Germany, except a certain privileged circle of persons and, after 1964, retired ones. This is one of the reasons why Willi Hennig never took into consideration to move to East Berlin or the GDR.

The Hennig family lived contently in the flat in Opitzstrasse. They enjoyed excursions to the forest area of "Grunewald" and the cultural life in West Berlin. Although Willi Hennig wore a necktie also at home (Figs. 35, 49, 53) this does not imply that he was not relaxed. His eldest son Wolfgang recalls only extremely few occasions on which he had seen his father without a necktie. He reports that this was nothing especially characteristic for his father but rather normal among academics at that time (Doc. 195). Willi Hennig wore a necktie even when he went out for an excursion with his sons (Fig. 40).

Willi and Irma Hennig loved Berlin, and they expected that the separation of Germany would reverse within their lifespan (Doc. 101). Consequently, Willi Hennig never seriously considered another position than

Fig. 40. Bernd, Gerd, and Willi Hennig in the Grunewald forest of Berlin in the 1950s, taken by Woldfgang Hennig. Courtesy of Wolfgang Hennig, Kranenburg.

one within the DEI. It is therefore entirely speculative to evaluate his chances of being appointed professor at the forestry department of the Technical University of Dresden in 1957. As it was usual in those times, the position had not been advertised, but Hennig received a letter from the dean of the forestry faculty, Rudolf Kleinert, in which he informed Hennig that he had been suggested as a candidate for a professorship, and he asked him to hand in a questionnaire on his personal data (Doc. 102), but Hennig never did. In 1959, he was nominated member of the Deutsche Akademie der Naturforscher Leopoldina—certainly the greatest distinction for a scientist available in Germany.

Hans Sachtleben turned 65—the age of regular retirement in Germany—in 1958 but continued acting as the director of the DEI (until 1962). However, at around the end of the fifties the Agricultural Academy sought for a possible successor. Willi Hennig had certainly good chances to get this position as he had a high scientific reputation, held the title of a professor, and was already head of a department, which means he had already reached the second highest rank in the institute's hierarchy.

According to several documents in Hennig's dossier at the intelligence service of the GDR ("Stasi", Doc. 103), he was indeed seriously considered if not secretly—but officially—accepted.

Possibly in preparation of his coming appointment, the intelligence service of the GDR decided to observe Willi Hennig. In July of 1959, the Stasi began to copy earlier documents and compiled them in a dossier. On June 19, 1959, one of Hennig's colleagues was interviewed by a Stasi-officer, Unterleutnant Güntzel. The latter suggested in his report (Doc. 103: 41) that his contact person should be hired as "inofficial contributor" ("Informeller Mitarbeiter": "IM", or "Geheimer [= secret] Informant": "GI") at the next occasion. This "secret informant" was Dr. Günther Petersen (Doc. 176), at that time member of the department of ecology and pest control at the DEI. He was extremely keen on writing reports (which he did under the alias "Christian"), and he feared—rightly so—that Willi Hennig would make efforts to enhance the taxonomic section of the DEI at the cost of the applied entomology department if he would become director. Petersen's reasoning was: The DEI is a public institution, based on financial support by the people of the GDR. Therefore, the budgeting should follow the "true interests" of the people, and this means that the scientists at the DEI should in the first place aim at enhancing the supply of the citizens with food and other natural resources. Consequently, the applied branch should be enlarged, whereas the purely taxonomic sectors had to be reduced and become subordinated under the directions of the applied ones. Someone who was honestly convinced of the socialist ideology could quite understandably be concerned about the future developments at DEI if a person with so little interest in applied sciences as Willi Hennig had decision making power. Petersen's fear was realistic since Hennig had pleaded for separating the applied entomology section from the remainder of the DEI and transfer it to Kleinmachnow, where the "Institut für Pflanzenschutzforschung der Akademie der Landwirtschaftswissenschaften der DDR" (Institute for research on plant protection of the Academy of Agricultural Sciences of the GDR) was in charge of plant protection and general agricultural research (Doc. 104).

The "Stasi" dossier provides an interesting insight into how the organisation operated. On one hand, they worked on an astoundingly unprofessional level. E.g., they noted (Doc. 103: 4) that it was not possible to ascertain where Willi Hennig worked. Also, they assumed that Hennig and his colleague Wolfgang Schwenke had founded a religious group amongst the DEI staff, obviously on no other ground than some papers of the "Evangelische Akademie" (in the GDR, nearly the only opportunity to be socially active outside the official socialist frame was offered by the two Christian confessions). Outstandingly absurd is the suspicion that Hennig had worked with IG Farben during WWII and was involved in the production of Zyklon-B gas which was used to kill jews and other people in concentration camps of the Nazis. A handwritten note to this point says "nicht bestätigt" ("not confirmed"), which is the least one could say (Doc. 103: 40). On the other hand, it is stunning how precisely "Christian" described what who said when to whom. He reported on discussions he had with staff members of Hennig's department, of things he had overheard, and he mentioned his own reservations against statements of Hennig and other staff. A major part plays the possible role of Willi Hennig in the escape ("Republikflucht") of Wolfgang Schwenke in 1959 (Doc. 103: 42–60). Although there were no hard facts, "Christian" as well as Unterleutnant Güntzel speculated that Hennig could have supported Schwenke's plans to leave the GDR. Formally, it was forbidden to leave the country without official permission. In practice, it was not difficult, at least not for single persons with light luggage: one had just to enter the local transport, especially the S-Bahn, in East Berlin and leave the coach in West Berlin. However, drawing larger amounts of money from an account in East-Germany, transferring a private library, or leaving jointly as a family required an elaborated logistics and could hardly be accomplished without help from other people.

The Stasi recognised without any reservation that Willi Hennig was an excellent scientist (e.g. Doc. 103: 16), but kept the—justified—suspicion that he did not really accept and support the new socialist political and social system. On p. 17 of the dossier we read that "the negative formation of the societal attitude of Dr. Hennig is proven already by the simple fact that he did not join the socialist workers' union" ("Die negative gesellschaftliche Bewußtseinsbildung der Herrn Dr. Hennig ist schon allein durch seinen Nichteintritt in den FDGB [Freier Deutscher Gewerkschaftsbund] bewiesen"). Originally, the ministry of agriculture and forestry of the GDR gave this judgment in connection with Hennig's appointment at the Potsdam College in 1951 (Doc. 206).

In a preliminary resume, dated 10.10.1960 (Doc. 103: 82), Leutnant Güntzel states that Willi Hennig had invited scientists from Eastern countries in his private home in West Berlin where he had defamed the political system of the GDR in a manner typical for the Western agitation in the press and on broadcast. Furthermore, Güntzel writes that Willi Hennig's salary was 3100.- DM (in these times "Deutsche Mark = DM" was also the valid currency in the GDR, however different from the Deutsche Mark in

West-Germany), of which one third was paid in "DM West". According to this report, Hennig used only a few percent of the remaining two thirds for shopping in the "democratic" (= East) sector of Berlin. "Paid in DM-West" is a euphemism, because the amount was paid by the East German academy in DM-East and exchanged at a rate of 1:1 by the Senate of West Berlin. The Senate supported by that tactic all those who earned their living in East Berlin while they lived in West Berlin. The Senate's interest was to maintain the idea of a united Berlin by encouraging practical connections between the two parts of the city. This means that the Hennig family had an amount of 1000 DM West to cover their costs in West Berlin. It is difficult to compare this amount to present days' purchasing power. But even if it might roughly be equivalent to 2000 €, it is striking that Willi Hennig should not have spent more DM East in order to lessen the strain on his DM West account. Indeed, as his eldest son, Wolfgang, remembers (Doc. 172) the Hennigs purchased whatever they could in East Berlin, certainly for more than only "a few percent" of Willi Hennig's East German money. This is entirely plausible because convenience goods were highly subsidised by the East German government and therefore much cheaper in East- than in West Berlin. The youngest son Gerd remembers that his mother went for an extensive shopping tour to East Berlin on a weekly basis, and that he often accompanied her and helped her carrying home the goods (personal communication 2012).

In the course of the following months, it became clearer and clearer that Willi Hennig could be advanced to director of the DEI after Sachtleben's retirement. At the same rate as this became more likely, the tone of "Christian"'s reports became more aggressive or even hysterical. He wrote minutes of several pages and made enormous efforts to convince his addressees that Hennig's directorate would mean a disaster for the DEI, especially for the ecological department, which was in his view the most important for the country's economy. He villainised vigorously the achievements of Hennig's department (Gerrit Friese, Eberhard Königsmann, Johann Machatschke) as "completely useless" (Doc. 103: 149–153). "Christian"'s efforts to marginalise Hennig and to prevent him from becoming director culminated in a furious memorandum (Doc. 103: 174–191), in which he argued that the tradition of the DEI as a home of basic research should be exterminated ("vernichtet"), the taxonomists should be forced to work as helpers for the applied branch, and the whole DEI should be restructured as an institute completely devoted to applied entomology.

Evidence of the infamy of the subversive efforts to suppress Hennig is an official inquiry of a colonel Weidauer at the head of the division

(Hauptabteilung) I of the Stasi, if Hennig's affiliation to the different military units during WWII allows inference on war criminal activites ("kriegsverbrecherische Tätigkeit", Doc. 103: 159). The answer by a major general Kleinjung (Doc. 103: 168) was short and clear: no.

Seen from outside, Willi Hennig had excellent prospects at the DEI. Admittedly, it was inconvenient to commute from his home in Steglitz to the DEI in Friedrichshagen (which took ca. 90 minutes by S-Bahn one way), but there he had an excellent library and collection at hand, a team of assistants and technicians, and almost certainly he would have advanced to director within a foreseeable time. Nevertheless, he must have thought of alternatives, as documented by a letter to Karl Strenzke of 05.03.1961 (Doc. 105). Strenzke had been a member of the Anti-Malaria unit at the Military Medical Academy from 1942 to 1945 (Steffan 1962), and Hennig knew him since his mission in Upper Italy, where they had met. Hennig replied to an inquiry of Strenzke for an advice because he (Strenzke) desperately looked for a new position, although he had a well-paid position as a researcher at a Max-Planck-Institute. Hennig wrote that if he had seen a halfway adequate field of activity ("ein halbwegs angemessenes Tätigkeitsfeld") elsewhere, he had used it for himself. In this revealing letter, Hennig mentioned that he permanently reckoned with the chance that he had to stay at home due to some events ("irgendwelche Ereignisse") from day to day ("von einem Tag zum anderen").

In July and August, 1961, Willi Hennig travelled through France, accompanied by his youngest son Gerd, for about four weeks, it was their summer vacation. They had made a round-trip through this country, and as usual, Willi Hennig had visited collections and colleagues on his way. When they headed for their return, they stayed in a pension at Colmar, near the German border, the night from August 12 to 13. When they started after breakfast, the landlady told them that in the morning of that day, something horrible had happened in Berlin, without giving details. Thus, they heard only in the evening, when they arrived at a relative's home in Mühlacker (ca. 40 km North-East of Stuttgart), about the erection of the Berlin wall (Doc. 106).

Non-Germans, and even young Germans, can hardly imagine what the wall meant to the citizens of Berlin in daily life, but even more so to their feelings. Still on June 15, 1961, Walter Ulbricht (then chief of the government of the GDR), had publicly stated "niemand hat die Absicht eine Mauer zu errichten" (nobody intends to erect a wall). Now, out of a sudden, families got torn apart, friends could not visit each other any longer, visitors could no longer move around freely. Of course, it was easier for

West Berliners than for people living in East Berlin or in the GDR. Anyway, Willi Hennig decided to quit his job in East Berlin immediately. He returned to Berlin on August 14, crossed the border to East Berlin several times by public transport and by car and transferred documents, specimens, private books and a dissecting microscope from Friedrichshagen to West Berlin.

He might have hoped that he could return to the DEI one day, but this would have been possible only more than a decade after he had died. Hennig was still on good terms with Hans Sachtleben, as shown by several letters the two of them wrote and received after August, 1961. There was some ambiguity on the question when exactly Hennig ceased working at DEI—he argued that he had still a number of leave days left, but on his staff record card at DEI the 13th of August, 1961 (Doc. 108), is given as his last working day. The avid rapporteur "Christian" gave a note to the Stasi on 18.10.1961 that he had heard that Hennig had "reportedly" ("angeblich") asked Sachtleben to keep his (Hennig's) position at the DEI vacant until January, 1962 (Doc. 103: 203). Be this as it may, Willi Hennig never returned to the DEI and had, as he had feared, to find an alternative source of living in the West. He would only have had the possibility to continue working at DEI if he would have moved to East Berlin (or the GDR), which would have meant living under unacceptable political and difficult economic conditions (Peters 1995). His decision to leave the DEI had far reaching consequences, as he was already envisaged as the future Director of the DEI. Quitting DEI cost him a number of privileges and opportunities, but Hennig was an ardent anti-communist as documented extensively by the Stasi (Doc. 103) and confirmed by his son Wolfgang (Doc. 101). For him, it was completely unimaginable to send his sons to a school or a university in socialist Germany.

2.7. *After the Wall—From Berlin to Ludwigsburg* (Fig. 41)

Life was complicated and uncertain for the Hennig family—and other scientists who had worked in East Berlin and left their working place as a consequence of the erection of the wall, e.g. Fritz Peus (professor at the Berlin university and director of the Zoological Museum), Stefan von Kéler (curator of entomology at the Zoological Museum), or Heinz Wermuth (curator of reptiles at the Zoological Museum). They all had lost their familiar working environment, their professional prospects, their loved field of work, and nobody knew for certain what would happen

Fig. 41. Willi Hennig at his microscope in the museum of Ludwigsburg, 1966. Courtesy of Gerd Hennig, Tübingen.

to their old-age pension and benefits. One of the—few—advantages of working in East Berlin and living in West Berlin had been the cheap medical care in East Germany and the favourable exchange rate between West and East German currency (1 DM West was officially 3 DM East, on the black market ca. 5 DM East).

Irma Hennig could buy certain food items and other things of daily need in East Berlin before the erection of the Berlin wall. However, for this purpose she needed a permit of the West Berlin customs authorities to pass the border, or rather to "import" the goods. This permit was withdrawn after August 13, 1961. Thus, in addition to all the trouble and confusion, life became more expensive, too.

In this situation, the governmental authorities in West Berlin and West Germany developed in a remarkably short time a programme to support those "refugees" ("Ostzonenflüchtlinge") who had left their positions in the East due to the wall. Willi Hennig could receive a short-term grant from the Deutsche Forschungsgemeinschaft on the basis of an application of less than 1½ pages (61 lines, Doc. 109), and in early December, 1961, he was employed as a professor at the Technical University of Berlin (West), on the basis of a retroactive contract from 1st of November on. The contract was valid preliminarily for two months but should have been extended until a permanent position—already envisaged at that time—for Hennig was established at this university (Doc. 110).

At Technische Universität Berlin, Hennig began teaching the first week of November, 1961, and gave two lectures and one practical course on invertebrate zoology (Schmitt 2002) until April, 1963. From a letter to Hans Sachtleben (Doc. 111) we see that the position of a full professor at a university did mean something to Willi Hennig. In addition, he could stay in Berlin, a place which he liked very much, for various reasons of which the strongest were most probably entirely emotional. Even under the difficult economic and political conditions of that time, Berlin was a cultural centre in Germany, yet being only West Berlin. Although Willi Hennig was not born in Berlin, he had taken a great fancy of this multifarious, vibrant city. As the public budget was substantially subsidised by the West German government, exhibitions, concerts, and theatres flourished and were affordable. Not to forget, here lived many acquaintances and close friends, above all Klaus Günther.

I find it astounding that several colleagues of Willi Hennig spontaneously tried to help him. Alfred Kaestner, who had left the Natural History museum of Berlin in 1957, and was professor at the University of München until 1967, alarmed the Hochschulverband (sort of professors' organisation in Germany) as they gave allowances to indigent colleagues like refugees from East Germany in a fast and unbureaucratic manner. Unrequestedly, he added a letter of support for Hennig. Finally, Hennig did not use the offer from the Hochschulverband as he got the position at the Technische Universität. In January, 1962, Elmo Hardy from the University of Hawaii at Honolulu expressed his concerns about the fate of Willi Hennig and his family (Doc. 112) and asked if Hennig could consider moving to Hawaii for a professorship at his own university. Elmo Hardy knew the Hennig family since his visit to the DEI in 1954 (Evenhuis 2004) and had become a real friend in the course of the subsequent years. He not just inquired about Hennig's condition but invoked Judson Linsley Gressitt, head of the entomology department at the Bishop Museum to offer a position as a dipterist at that museum (Doc. 113), and made certain that the offer was made officially (Doc. 114) by the director. Hennig declined (Doc. 115), but emphasised that he certainly would have liked to co-operate with Linsley Gressitt (whom he had met in London in 1958) and especially with Elmo Hardy whom he called a friend. The reasons he gave for his decline were the great distance from Honolulu to the European insect collections, which he needed to visit for the continuation of his "Larvenformen der Dipteren", and the education of his three sons, who were aged 17, 19, and 21 at that time. On one hand, he did not want to take them out of their academic and social environment; on the other hand, he did not want to leave them in Germany, as he had to live under very provisional conditions during the first ten years of their family life. Likewise, he did not accept an offer to go "for some years" to the Entomology Research Division at the U.S. Department of Agriculture in Washington, D.C. (Doc. 089).

Already early in 1962, there must have been negotiations with the Staatliches Museum für Naturkunde, Stuttgart (SMNS—State Museum of Natural History), as Willi Hennig signed their questionnaire on 11.03.1962 (Doc. 037). As a consequence of the Berlin wall, the federal government had developed a support plan for those scientist who lost their positions in East Berlin or East Germany. Partly from these funds, a department could be opened at the SMNS, dedicated to Phylogenetic Research, ad personam Willi Hennig, due to the activity of Ernst Schüz, director of the SMNS from 1949 to 1969 (Ziegler 1991). This department was housed provisionally in the city of Ludwigsburg, 15 km north of Stuttgart (Fig. 42).

Fig. 42. Ludwigsburg, Arsenalplatz 3, the former SMNS. Willi Hennig's office was on the second floor and had two windows, hidden by the tree on the right. Original, taken 29.10.2012.

Willi Hennig had outlined what he thought could be the configuration of such a department (Doc. 116). He suggested that he should have three scientific assistants, a secretary, and at least one technician. His demands for technical equipment were remarkably modest: one microtome and one thermostat and for each scientist one light microscope and one dissecting microscope. Already in this first version, he admitted that it would not be possible for financial reasons to hire three assistants, so he suggested—in a very moderate tone—to start with one.

Willi Hennig started working at the SMNS on April 1, 1963, and the Hennig family moved into a flat with four rooms at Denkendorfer Str. 16 in Ludwigsburg-Pflugfelden. The entire transport of furniture, books, equipment and other items and of the members of the family had to be organised by plane, because Hennig's name was on a wanted list of the East German police so that he would have been arrested if he had shown up at the border (Doc. 201).

From their new home, the distance to the museum at Arsenalplatz 3 was ca. 3 km which Hennig daily covered by car. Normally, he arrived at work around nine o'clock. He brought sandwiches in a box (and expected his technician to prepare a cup of tea at noon), so that he needed not to leave his office for lunch (Doc. 153).

Indeed, he had to be content with only one assistant in the beginning. This was from July 1, 1963 until 1966 Hans Ulrich, followed on 01.01.1967 by Dieter Schlee, and only in 1970 a second assistant—Wolfgang Seeger—and a second technician could be hired (Doc. 117, Doc. 118). Immediately after starting work at SMNS, he began to publish in the museum's journal "Stuttgarter Beiträge zur Naturkunde"—of the 44 publications from 1963 until his sudden death in 1976, 29 appeared there. Of course, many of his papers dealt with Diptera taxonomy and systematics, but the focus slightly shifted to amber fossils. Dieter Schlee contributed much to the growing importance of this field of studies. Hennig had no curatorial duties, he was not responsible for a part of the museum's insect collections, but he regularly used them (see Fig. 1).

The new institution clearly offered considerably more chances for Willi Hennig to visit museums abroad, to attend congresses and to meet colleagues. According to a draft of a department's report on the years 1963 and 1964 (Doc. 118), he joined the annual meeting of the German Zoological Society at München in 1963, the so-called Marburg Symposium on "problems of polyphyly" (29.11.–01.12.1963), the 12th International Congress of Entomology in London (July 6–16, 1964), the 1st International Congress of Parasitology in Rome (September 21–26, 1964), and a symposium on systematics and evolution at the university of Freiburg im Breisgau

(November 15–16, 1964). Especially the International Congress of Entomology in London was a spledid opportunity to meet colleagues from abroad who knew Hennig's name and whom he knew through correspondence and publications (Fig. 43).

In addition, he visited the museums of Helsinki (Finland), Stockholm (Sweden), Copenhagen (Denmark), London (England), Paris (France), Milano and Florence (Italy) to see their Diptera collections. Interestingly, he mentions that he was at that time working on a manuscript on Phylogenetic Systematics destined for Annual Review of Entomology on request of its editors. The reason Hennig attended these and other meetings was not in the first place to sit in an audience and listen to lectures, but to find the opportunity for a chat with colleagues in a more private frame, and to visit collections in the vicinity of the venue. In June of 1963, e.g., he visited the famous Georg Frey collection at Tutzing, together with Klaus Günther (Fig. 44), after the annual meeting of the German Zoological Society.

Fig. 43. 12th International Congress of Entomology, London 1964, from left to right: Curtis Saborsky, (Robert L. Usinger, hidden,) Boris Rohdendorf, Willi Hennig, Norman Denbigh Riley. Courtesy of Gerd Hennig, Tübingen.

Fig. 44. Willi Hennig and Klaus Günther (right) in the library of the Museum Georg Frey, Tutzing, June, 1963. Courtesy of Waltraut Günther (1985).

2.8. *Early Debates* (Fig. 45)

The symposium on polyphyly in Marburg 1963 had a special impact on Willi Hennig. It was the 8th Phylogenetisches Symposium (Kraus 1984), with contributions by Walter Gross (palaeontologist from Tübingen, formerly Berlin), Adolf Remane (zoologist from Kiel), Peter Ax (zoologist from Göttingen), and Wolf Herre (zoologist from Kiel). Hennig was especially urged to attend by Wolf Herre, who had visited the Stuttgart museum and on that occasion also Hennig. Actually, Hennig had generally refrained from attending scientific meetings but could not resist Herre's pressure (Doc. 119). Thus, he went to Marburg and met there for the first time with Adolf Remane. Remane was the doyen of phylogenetic and evolutionary research in the German-speaking world and had published a most influential book on the "*Grundlagen des natürlichen Systems, der vergleichenden Anatomie und der Phylogenetik: theoretische Morphologie und Systematik*" (foundations of natural system, comparative anatomy, and phylogenetics:

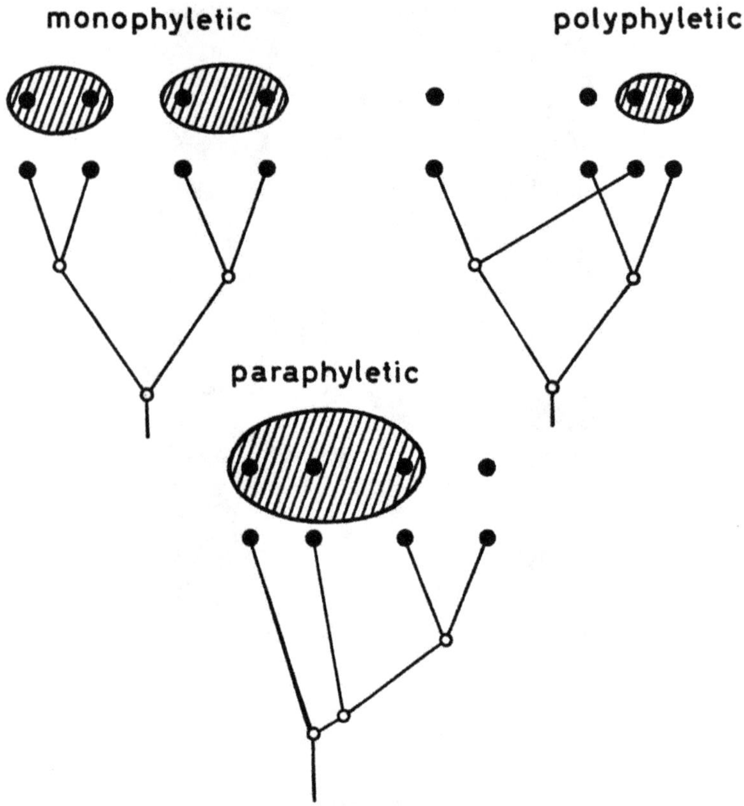

Fig. 2. The three different categories of systematic group formations corresponding to the resemblance of their constituents resting on synapomorphy (monophyletic groups), convergence (polyphyletic groups), or symplesiomorphy (paraphyletic groups). For comparison with Figure 1.

Fig. 45. Fig. 2 from *Phylogenetic Systematics* (1966).

theoretical morphology and systematics). In addition, he saw renowned palaeontologists such as Josef Kälin (Fribourg, Switzerland), Walter Georg Kühne (Berlin), Erich Thenius (Wien, Austria), and zoologists—Alfred Kaestner (München) and Curt Kosswig (Hamburg). However, he found the symposium little inspiring and would have regretted the journey to Marburg if there would not have been the subsequent extension to Denmark and Sweden.

The proceedings of the symposium were published in volume 173 of Zoologischer Anzeiger in late 1964 (including a short discussion remark by Willi Hennig on p. 63, no. 113a). When Hennig read the elaborated contributions, he must have been upset. Immediately, he jotted down a critique of 15 pages and sent it to Klaus Günther, asking for comments (Doc. 120). It took some time until Günther returned the annotated manuscript (Doc. 121), but he was highly enthusiastic about it and encouraged Hennig to proceed in publishing it. The two of them agreed that the symposium speakers, especially Gross and Remane, had no clear understanding of phylogenetics, that they used vague concepts, and did not consistently found their conclusions. The focus of their critique is that the authors used both terms, 'monophyly' and 'polyphyly', in an ambiguous and therefore fruitless manner. I found several and partly incomplete versions of this manuscript in Willi Hennig's bequest, kept by Gerd Hennig (Tübingen). Finally, Hennig and Günther agreed in giving up the project, mainly because a fundamental paper by Willi Hennig was released from the press in March, 1965.

This paper is entitled "Phylogenetic Systematics" and appeared in Annual Review of Entomology vol. 10. Its genesis throws some interesting light on the state of zoology (and perhaps science in general) in Germany in the early 1960s. In summer or early autumn of 1963, the committee preparing the decennial volume of Annual Review of Entomology (P.A. Dahm, V.G. Dethier, G.F. Edmunds, L.E. Rozeboom, B.N. Smallman R.F. Smith, and E.A. Steinhaus) had asked Willi Hennig to contribute a concise paper of 40 pages in which he could outline his method to a broader, English-speaking community. Deadline for submission of the manuscript was April 1, 1964. Hennig had prepared a German version until December 30, 1963 and had sent it to his friend Klaus Günther (Doc. 122), whom he asked in the same letter if he knew someone who could translate the text into English. He could offer US$ 50, the amount the editors of Annual Review of Entomology had announced to provide. Günther was excited about the manuscript, called it "a true masterpiece" ("ein wahres Meisterwerk"), and promised to ask a US-American student of medicine for the translation (Doc. 123). Shortly after, Hennig asked his colleague Joachim Illies if he could advice him on that matter: "... according to the request of the editors, I should care for a translation into American myself. With that I have some problems. I do not want to translate the text myself. This costs me too much time, and in addition I do not feel certain enough" ("... nach dem Wunsche der Redaktion eine Übersetzung ins Amerikanische selbst anfertigen lassen soll. Damit habe ich nun einige Schwierigkeiten. Selbst möchte ich die Übersetzung nicht machen. Das kostet mich zuviel Zeit und außerdem fühle ich mich nicht sicher genug", Doc. 124).

From today's perspective, it might appear strange that two highly educated, experienced, and renowned German zoologists did not dare to translate a manuscript of less than 40 pages (the printed version is only 20 pages long) into English. They did not even consider preparing a preliminary English version and then asking a Native speaker, e.g. one of their Anglo-Saxon colleagues, to review it. However, they were definitely not the only ones in their community, which means that in those times German scientist (at least zoologists) could confine themselves to communicate their results in German exclusively. However, even in those times it was clear that someone publishing exclusively in German was hardly recognised outside the German-speaking community. This applied not only to Hennig and Günther, but also to Adolf Remane, Gerhard Heberer, and many more scientists who were well known and highly respected in German-speaking Central Europe (Schmitt 2001: 334).

Willi Hennig found a solution himself. He asked a British dipterist, Graham Charles D. Griffiths, whom he knew from publications and whom he estimated for his phylogenetic skills (Doc. 125), if he would be willing to translate the paper, and Griffiths agreed in an extremely kind manner on February 7, 1964 (Doc. 126). He completed the translation until April, 1964 (Doc. 127), and Willi Hennig could submit his manuscript right in time. Hennig received the reprints of "Phylogenetic Systematics" on March 8, 1965 (Doc. 127). This paper was the first of Hennig's to appear originally in English. To date, it is worth reading as a concise and well-written overview on his method.

Things changed crucially when in winter of 1966 the book *Phylogenetic Systematics* was released from the press. Willi Hennig had written a widely revised version of his *Grundzüge* (which had been called "the most important nonavailable book during the past two decades for English-speaking systematists" by Walter J. Bock in 1968). As Hennig emphasised repeatedly (e.g. Doc. 129), he had submitted his book manuscript in 1961 (it is unclear whether before or after August 13), but had no news about the progress of the translation, and that he had not seen the English text until the work was published. As discussed in more detail below (chapter 5.2.), the English book is at any rate an adequate representation of Hennig's ideas. It was recognised immediately in the English-speaking scientific world and marks the actual starting point of the "Hennigian revolution" (e.g. Dupuis 1990; Mishler 2000; Wheeler 2008).

Between Hennig and Günther, however, *Phylogenetic Systematics* was no topic. During several months from mid-1967 until late 1969, the correspondence between them (e.g. Doc. 130, Doc. 131) was dominated by their joint concern about Hennig's colleague Gerd von Wahlert, who had joined the museum staff at Ludwigsburg in 1962 as an ichthyologist, coming from the institute for ocean research (Institut für Meeresforschung) in Bremerhaven. Gerd von Wahlert had managed to receive a specimen of *Latimeria chalumnae* in 1966 and planned to produce a book on *Latimeria* and the evolution of the vertebrates as an accomplishment to receive the 'habilitation'. The 'habilitation' is a typical Central European (or German) section of an academic career. Any scientist who aimed (or still aims, as approaches to eliminate this instrument are still not brought to an end) at becoming a professor had to pass a last academic examination. One requirement was the production of a 'Habilitationsschrift', i.e. a comprehensive publication (other prerequisites are teaching experience and an excellent publication record). Before the reformation of the universities in Germany in 1969/1970, the decision on a 'habilitation' project was made by the entire faculty, which meant in the first place: by the full professors. Consequently, scientific aspects were not the only ones playing a part in the procedure. In the case of Gerd von Wahlert's intention to habilitate, a whole number of scientific, political, psychological, and social reasons caused complications and delays. He worked at a museum, so he had to find a university to go. His views on vertebrate evolution were unconventional, so the scientific establishment was extremely reluctant to grant the honour of 'habilitation' to it. Within the museum, non-habilitated colleagues resented von Wahlert the formally higher academic rank. To be habilitated means inevitably a teaching duty, which means less time for ordinary museum work. One who gets habilitated is always under the suspicion of seeking to leave the museum and get a professorship at a university. Which of these possible factors were how important in Gerd von Wahlert's 'habilitation' is irrelevant here. At any rate, Hennig and Günther were intensively involved in this matter. They discussed with von Wahlert and sought advice from each other. Finally, Gerd von Wahlert received the 'habilitation' from the Freie Universität Berlin with the massive help from Klaus Günther in 1969 (Doc. 132).

2.9. *Attempts to Improve a Situation* (Fig. 46)

Werner Ulrich, full professor of zoology at the Freie Universität Berlin (West), announced in autumn, 1965, to retire in April, 1966. The faculty of mathematics and sciences formed a committee to find a successor. The committee's head was Klaus Günther, one other member was the palaeontologist Walter Georg Kühne (see box 5). Kühne had met Hennig in the 1950s already (Kohring & Schlüter 1997) and admired or even adored him.

Fig. 46. Klaus Günther handing over the diploma of the honorary doctorate to Willi Hennig on 21.02.1969 in Stuttgart. Courtesy of Waltraut Günther (1985).

Box 5 Walter Georg Kühne

"One of the most colourful ['schillernd'] figures in German palaeontological research" (Schlüter 1981) was born on 26.02.1911 in Berlin as the second child of the artist Walter Kühne and Renata, née von Stülpnagel. He had an elder sister, Maria, and a younger brother, Wolfgang. He attended school at different locations and passed the final examination (Abitur) in 1930 at the *Deutsche Oberschule am Meer* on the North-Sea island of Juist. From 1930 he studied geology and palaeontology at the university of Berlin for three semesters and moved to Halle an der Saale in 1931. Due to his political engagement on the extreme Left, he was expelled from the university on October 6, 1933, and taken into pre-trial custody for nine months. In 1934, he married Charlotte Petsche (they divorced in May, 1956). For several years, he earned his living as a free-lancer, collected and sold fossil insects, and compiled a catalogue of medieval bells ornamented with a strand relief (paid by the Prussian General Conservator).

Towards the end of 1938, Walter Georg and Charlotte Kühne emigrated to London. In Britain, too, he made his living from selling fossils, until he received a position as a lecturer at the University College of London in 1944. He worked at that institution until 1951 and wrote his doctoral dissertation during this time. He submitted his dissertation to the University of Bonn, from where he received his doctoral degree in 1949. In 1952, he returned to Germany and went to Berlin, where he lived for the rest of his life. Again, he had to sell fossils to earn some money, but he also held temporary lectureships at the university, then in the Soviet sector of Berlin. In 1955, he became docent at the Freie Universität Berlin (the university founded in 1948 in West Berlin), where he 'habilitated' (qualified for professorship) in 1958, and was appointed associate professor in 1963, and full professor in 1966. He married Ursula Kühne, a distantly related cousin, in 1963. In 1976, he retired, and died on March 16, 1991 (all biographical data from Kohring & Schlüter 1997) (Fig. 81).

His main contribution to palaeontology is certainly his research on mesozoic mammals. His most spectacular publication is a 130-pages-book on palaeontology and dialectic materialism, which appeared in East Germany in 1979. He had met Klaus Günther in 1951 who arranged a personal contact of Kühne to Willi Hennig. Kühne was doubtless the first, and for decades the only, German palaeontologist who accepted Hennig's phylogenetic systematics. Kühne's ardour for Hennig's method can clearly be seen from the enthusiastic tone of his letters to Hennig in the 1960s and 1970s. He wrote an emphatic appreciation of Hennig's contribution to science, in which he stated that Hennig's *Grundzüge* caused a revolution in systematics (1978). Walter Georg Kühne was "a rebel who spoke his mind, right or wrong", and he will be remembered "as an inspiring and thoughtful palaeontologist" by those who knew him personally (Kohring 1991).

To illustrate how remarkable his personality was, I describe two events which I witnessed myself. At the 21st Phylogenetisches Symposium in Göttingen in 1976 on the phylogeny of the arthropods, Kühne sat in the second row of seats, just behind Rolf Siewing (student of Adolf Remane and professor of zoology at the University of Erlangen), when the audience discussed lively on "typology". Kühne bent forward and said something to Siewing, whose face immediately turned alarmingly red. When returning from the following coffee break, Siewing approached Peter Ax and complained highly upset about Kühne. He shouted: "Can you imagine what Kühne said to me? He said that 'typologist' is in his view the dirtiest word for a scientist!" ("das schlimmste Schimpfwort für einen Wissenschaftler"). Exactly that moment, Kühne came round the corner and had obviously heard Siewing's last words. He grinned at Siewing, wagged his finger at him and said "Yes, yes, Herr Siewing". By the way, Kühne was—in my opinion—completely right: Siewing was the archetype of a typologically thinking zoologist.

The topic of the 35th Phylogenetisches Symposium in 1993, held at the Freie Universität Berlin (West), was the species concept. In a coffee break, I talked to Kühne at the base of the stairs in the institute building. Roughly 100 coffee drinking and chatting attendants populated the hall and the staircase. We agreed that all the preceding discussions on Hennig's approach to delineate species in time were completely superficial and in vain. Kühne drew breath and spoke with stentorian voice "Exactly, the only relevant question is here: when has who with whom fucked the last time" ("Sehr richtig, die einzig relevante Frage ist doch WANN HAT WER MIT WEM ZULETZT GEFICKT!"). The conversation in the hall and in the staircase stopped abruptly, and Kühne looked at me with a whimsical smile, obviously enjoying the public shock he had caused.

Fig. 81. Walter Georg Kühne in 1987, during his journey in New Zealand. Courtesy of Rolf Kohring, Berlin.

He sent amber fossils to Hennig, and exchanged publications with him. On November 30, 1965, he wrote a fervid letter to Willi Hennig in which he indiscreetly reported from the meeting of the committee. Other members of the committee had mentioned Georg Kümmel as a candidate for the position vacant in the coming year. Kühne had protested and claimed instead that "for Berlin we need someone with an international standing" ("für Berlin brauchen wir einen mit internationaler Wertschätzung"). He asked Hennig to send as soon as possible names of highly renowned zoologists who would agree to nominate him, and reviews of his publications, so that he had arguments for the discussions in the committee (Doc. 133).

Willi Hennig wrote an extensive letter to Klaus Günther (Doc. 134) in which he described in all detail his situation at the museum in Ludwigsburg. He mentioned Kühne's letter, but also an informal inquiry about his willingness to accept a professorship at the University of Hamburg. From this and following letters it can clearly be seen that Hennig would have hesitated to leave his post in Ludwigsburg, but also that he would have considered seriously to return to Berlin and to work at a university if there would be an attractive offer. Günther replied surprisingly open (Doc. 135). He explained in length why Georg Kümmel would be a perfect candidate: he had a broad and profound knowledge of morphology, physiology, and ecology of a wide array of animals, he was a specialist on the then new technique of transmission electron microscopy, and he was also trained in philosophy and psychology. Thus, to nominate Hennig would certainly not lead to his appointment, but could possibly yield a means to enhance his status at the Ludwigsburg museum. To put his name on a short list of a professorship appointment would definitely underpin his reputation in the academic community. Consequently, Günther did not really fight for an appointment of Willi Hennig as professor at the Freie Universität Berlin, but tried hard to get his name on the short list (Doc. 136).

Obviously, Walter Georg Kühne did not share Günther's view. He wrote in haste on November 5, 1966, that the committee had listed Hennig "primo et aequo loco" together with Georg Kümmel, and argued that only one vote was missing or Hennig's name would have been the first (Doc. 137). Just the day before, Klaus Günther had sent a handwritten letter of eight pages in which he, among other issues, reported on the decision of the committee—W. Hennig and G. Kümmel jointly on position one, M. Renner on position two ("primo et aequo loco W. Hennig, G. Kümmel, secundo loco M. Renner", Doc. 138). He pointed out—realisticly—that on a higher political level "probably Kümmel will be chosen, simply because he is younger" ("weil man höheren Ortes vermutlich auf den Kümmel gehen wird, schon weil jünger"). Thus, Hennig must have gained a rather adequate picture of his chances in Berlin, but not so Kühne. Günther described vividly how the members of the faculty's council discussed on different options, how Kühne fervently ("feurig") pleaded for Hennig, and that finally they decided with 31 against 14 votes to list Hennig second, after Kümmel and before Renner (Doc. 139).

Walter Georg Kühne was certainly a misfit in the faculty, and he knew it. He wrote to Hennig that "given my status at the faculty, every case which I make to my own is doomed to get rejected" ("Bei meiner Stellung in der Fakultät ist jeder Fall, den ich zu meinem eigenen mache, der Ablehnung unterworfen", Doc. 140). Nevertheless, he made another attempt to help his hero. In February, 1969, he asked six prestigious professors—Peter Ax (Göttingen, Germany), Lars Brundin (Stockholm, Sweden), Joachim Illies (Schlitz, Germany), Otto Kraus (Frankfurt am Main, Germany), Günther Osche (Freiburg im Breisgau, Germany), and Klaus Günther—to support an application to the Max-Planck-Gesellschaft to open a research institute for Willi Hennig (Doc. 141). The background of such an idea was the fact that the DEI had been run by the Kaiser-Wilhelm-Gesellschaft, which was re-founded and renamed after WWII as Max-Planck-Gesellschaft. Otto Kraus and Günther Osche replied immediately (Doc. 142, Doc. 143), the two of them emphasised their unconfined high esteem of Willi Hennig but expressed at the same time a certain scepticism regarding the chances of such an attempt. Their main points were that Willi Hennig was aged 56 by that time, so he had only nine years of professional activity left, of which he had to invest a considerable portion in establishing a new Max-Planck-Institute from the scratch, and that they saw absolutely no chances to build an insect collection at a newly founded institute that could be comparable to the one at the Ludwigsburg museum. Hennig read all these papers, and he wrote an extensive letter to Kühne (Doc. 144) in which he unmistakably stated that he saw hardly any chances that the Max-Planck-Gesellschaft could accept Kühne's suggestion. The Max-Planck-Gesellschaft fostered experimental branches of biology exclusively, and indeed, they finally disapproved Kühne's application, lastly arguing that their financial means were not sufficient (Doc. 145). Although Hennig had stated he had no hope, his discussion of Kühne's suggestion (Doc. 144) gives the clear impression that he would have appreciated an Institute of Phylogenetic Research financed by the Max-Planck-Gesellschaft. He lamented his situation at the museum at length: He was dependent in nearly all respect on the good will of the museum's director, he had no budget on his own, due to the lacking interaction between the museum and any university he had no doctoral students, and he had only one assistant instead of three (as he originally had hoped). This is clear evidence that he would not have refused to become head of such a Max-Planck-Institute, in spite of the initial non-scientific workload. After all, he argued, could the Max-Planck-Gesellschaft possibly just pay for some additional staff and give him an individual budget for his research but let the museum still allocate rooms, collections, library, and administrative infrastructure. Considering the peculiarities of the German science business (due to the federal structure of Germany), the pettishness of any administration when it comes to cuts in budgets and competences, and—not least—the impending envy of colleagues of other departments, one could easily anticipate that this plan had never any prospects.

Hennig's friend Klaus Günther followed a different path in order to support him, and he succeeded. He proposed to the faculty of Mathematics and Sciences of the Freie Universität Berlin to award an honorary doctorate to Willi Hennig. Until then, only one biologist had been honoured that way, Richard Goldschmidt in 1953 (Satzinger 2009: 194). On December 4, 1968, the faculty approved the application by eight biologists on the bestowal (Doc. 146), and the dean signed the certificate the same day. Willi Hennig recognised the honour with deep gratitude but asked the colleagues in Berlin to apologise that he could not travel to Berlin to receive the honorary doctorate in the course of an official ceremony. He mentioned as the reason of his decision that he suffered from an insufficiency of his heart and blood circulation (Doc. 147) which prevented him from longer trips. Finally, Günther came to Ludwigsburg as a replacement of the faculty dean and handed the certificate over to Hennig on March 21, 1969, during a ceremonial act at the guesthouse of the factory owner Dr. K.-E. Scheufelen in Stuttgart (Fig. 46). In his speech, Günther emphasised Hennig's merits as the founder of Phylogenetic Systematics, but also as a taxonomist, and he flattered the museum representatives for their great luck that a luminary like Willi Hennig was a member of their staff (Schüz 1969). Reading this laudation, one cannot avoid the impression that Günther had chosen his wording in order to impress on all the decision makers from museum and politics so that they would get more willing to fulfil Hennig's desires. Perhaps this manoeuvre contributed a bit to Hennig's possibility to hire additional staff in 1970 (Doc. 117, Doc. 118).

Hennig kept the title of a "professor" since his appointment at the Potsdam college, but he did not teach at a university in West-Germany until 1971, and seemingly he did not supervise diploma- or doctoral students. Already during the early 1960s, as soon as he had moved to Ludwigsburg, he established contacts to the University of Tübingen. Here, an acquaintance of his from war times had a position as a professor of zoology—Karl Grell who had done service in the same anti-malaria unit as Willi Hennig. He visited Grell several times and got also into contact with Gerhard Mickoleit (Doc. 149, Doc. 175).

On February 27, 1970, Hennig was appointed honorary professor at the University of Tübingen, after a case history entirely unusual in the German science scene. The idea to appoint him was born in student circles (Doc. 148). During a student strike in 1968, several students sat vividly discussing in the clubhouse of the University of Tübingen, when one of them—nobody can remember who it was—around three o'clock in the morning brought up the plan to entice Willi Hennig to lecture at the Tübingen University. These students had heard of Hennig's method through Gerhard Mickoleit in a seminar during the advanced course in comparative animal anatomy (Doc. 149). The students' representative in the faculty council, Hans-Jörg Mayer, checked how to proceed and submitted the formal proposal to the council. The dean at that time, palaeontologist Prof.Dr. Adolf Seilacher, forwarded the proposal to the rector on October 31, 1969 (Doc. 149).

Beforehand, however, two students—Helmut Schmalfuß and Helmut Oelschläger—had travelled from Tübingen to Ludwigsburg as delegates of the student community and visited Willi Hennig at the museum. They wanted to sound out Hennig on the chances that he would accept the honorary professorship. The two of them came unannounced to Hennig's office. They met the admired phylogeneticist as he sat at his desk, playing with a cold cigarette (at times, he smoked a lot but tried intermittently to smoke less or stop smoking at all, and spinning cold cigarettes with his fingers was a typical attitude of his when he reasoned intensively on something—Doc. 172). Hennig was deeply impressed by the desire of the students, since after the announcement by the gate keeper he had obviously expected to see visitors who wanted to get some six-legged critters identified (Doc. 148).

Joachim Illies, with whom Willi Hennig cultivated a friendly correspondence, congratulated him on March 25, 1969, for the honorary doctorate from Berlin and mentioned that he had heard the rumour that Hennig had been appointed professor at Tübingen (Doc. 150). Already the next day, Hennig replied (Doc. 151): "highly astonished I read in your letter what you write about Tübingen. After some hard thinking I ended up presuming that your (unfounded) information must be connected with the student Schmalfuß. This guy was here (together with a fellow student) and hinted on the circumstance that some students were dissatisfied by the lack of a lecture on Special Zoology (and pertinent questions) at Tübingen, and that they wanted to put things right and that they probably had thought of me somehow" ("Mit großem Erstaunen las ich in Ihrem Briefe, was Sie von Tübingen schreiben. Nach einigem Kopfzerbrechen bin ich zu der Vermutung gekommen, dass Ihre (nicht zutreffende) Informationen mit dem Studenten Schmalfuß zusammenhängen müssen. Der war vor einiger Zeit hier (mit einem Kommilitonen) und ließ durchblicken, dass einige Studenten mit dem Fehlen einer Vorlesung über Spezielle Zoologie (und einschlägiger Fragen) in Tübingen unzufrieden seien und versuchen wollten, Abhilfe zu schaffen, wobei sie wohl in irgendeiner Form an mich gedacht hatten").

In the semester following his appointment (summer of 1970), Hennig asked for dispensation from his teaching duties (the university expected him to teach for two hours per week each semester). But after that he gave four times seminars of two hours each in which certain animal taxa were extensively treated from a phylogenetic perspective. In winter 1970/71, the seminar was on general topics in systematics and phylogenetics ("Systematisch-phylogenetische Fragen"), in summer of 1971 the seminar was entitled "seminar on open questions in the phylogenetics of the animal phyla", in the winter semester 1972/73, Hennig gave the seminar jointly with Gerhard Mickoleit (on the phylogeny of the mammals with special emphasis on the primates), and in summer semester, 1974, they discussed on the systematic relationships of the molluscs (Doc. 152). As usual, the students' task was to report on relevant literature, give summaries, and thus contribute to the professor's book projects. These seminars were, in a way, test runs for coming editions of Hennig's pocket-books on zoology. During his time as an honorary professor at the university of Tübingen, Willi Hennig supervised—alone or jointly with a colleague—six doctoral students: Helmut Schmalfuß, Ernst-Gerhard Burmeister, Christian Rieger (supervised by Karl Grell. Hennig was co-supervisor), Werner Bils, Till Osten, and Martin Baehr (whom Gerhard Mickoleit adopted after Hennig's death in 1976). Hennig's correspondence kept in the archive of the SMNS clearly shows that he did not invest much time in his students. Only one of them, Till Osten, exchanged more than one letter with Hennig. Till Osten was presumably the only one who came several times during his doctoral project to Ludwigsburg and discussed its progress with Willi Hennig (T. Osten, personal communication 2000).

2.10. *Zenith and Sudden End* (Fig. 47)

At the museum, Hennig followed his daily routine, investigated on insect morphology and systematics, and focussed more and more on amber fossils. Between 1964 and 1977, 18 of his 39 publications were devoted to fossil Diptera in amber. In addition, he described new genera and species and contributed the 337-pages-chapter on Diptera to the *Handbuch der Zoologie* (published 1973)—all this "normal" science in Kuhn's (1962) sense throughout. Of special importance for the history of systematics are two of the publications after *Phylogenetic Systematics*: *Die Stammesgeschichte der Insekten* in 1969—a book of 436 pages—and "Kritische Bemerkungen zur Frage 'cladistic analysis or cladistic classification'" in 1974, a reply to a criticism of Phylogenetic Systematics by Ernst Mayr. These are discussed below (in chapter 5.3 and 5.4. respectively).

Willi Hennig corresponded with numerous colleagues; most of the correspondence kept in the archive of the SMNS pertains to Diptera, either

Fig. 47. Willi Hennig on August 12, 1976 (possibly the last portrait photo). Courtesy of Gerd Hennig, Tübingen.

loan of material from other museum collections or special questions on certain characters. Of these "dipterological" letters are of special interest those written to and received from Lars Zakarias Brundin, with whom Hennig agreed completely in the field of methodology. They donated each other their large and expensive books, they sent each other manuscripts for checking, and they discussed intensively on conceptual and terminological problems. Brundin played certainly a most important part in the spreading of Hennig's principles (e.g. Doc. 158). According to Hull (1989: 5), Brundin "set out his version of Hennig's phylogenetic system clearly and energetically". For the "evolution of phylogenetic systematics" (Hull 1989) it was a turning point when Gareth J. Nelson visited the Swedish Museum of Natural History in 1966 and learned through Lars Brundin of Hennig's approach. Nelson was soon "convinced of the superiority of Hennig's methodology to the principles espoused by Simpson, Mayr, and company. It was chiefly Nelson who served as the 'initial irritant' in bringing phylogenetic systematics to the attention of the biologists working at the British Museum (Natural History) and the American Museum of Natural History". The crucial factor in this story is that Brundin spoke and wrote German fluently, while Hennig had serious problems with expressing his thoughts in English and possibly also with understanding others' thoughts when expressed in English. The complete correspondence between Hennig and Brundin is in German. They have met personally at least three times—in June, 1968, when Brundin and his wife travelled by car to South Europe and stopped by at Ludwigsburg, in July, 1969, when he visited the museum at Ludwigsburg, and in 1972 during the 14th International Congress of Entomology in Canberra (August 22–30). Before the congress, they addressed each other as "Lieber Kollege..." (dear colleague) or "Sehr verehrter Kollege..." (honoured colleague...), after their Australian encounter they wrote "Lieber Lars" (dear Lars) and "Lieber Willi". Seemingly, they found also a warm personal relationship.

Gareth Nelson, at that time at the American Museum of Natural History (AMNH), did not only accept Hennig's methodology and applied it himself, but he also made considerable efforts to propagate it in which way he ever could. In 1971, Nelson had published a paper on "'Cladism' as a philosophy of classification" which is an eloquent defence of Willi Hennig against the critique by Darlington, Mayr, and Ashlock. Especially when he was on the editorial board of *Systematic Zoology* (from 1971–1973; 1974–1976 he was editor-in-chief) he managed to place several of his own statements in support of the Hennigian method. In addition, he succeeded in getting Hennig elected as a Corresponding Member of the AMNH (Doc. 159), he invited Hennig to join the editorial board of *Systematic Zoology* (Doc. 160), and he convinced him to agree in publishing an English translation of his reply to Ernst Mayr (Doc. 161). When Hennig agreed to review manuscripts for *Systematic Zoology* (Doc. 160), he symptomatically asked Nelson for permission to use German (he wrote all letters to Nelson in German).

Although Willi Hennig was not keen at all to give public lectures or to discuss in a group of more than four people, he liked travelling, and he continued attending congresses (in the first place to talk to colleagues personally and in a smaller circle of people—Doc. 172). His favourite destinations for private vacations were Greece, Italy, and Jugoslavia. He was especially fond of the Italian Renaissance and of the ancient Greek roots of western culture, and he had acquired quite an impressive command of Italian in the course of his journeys (Doc. 172). However, "most important for him were the scientific collections and the colleagues there" (Doc. 003).

In 1967, Willi Hennig was invited to stay at the Entomology Research Institute, Research Branch, Department of Agriculture, in Ottawa, Canada, most probably on an initiative of J. Frank McAlpine, the editor of the *Manual of Nearctic Diptera*. He worked in Ottawa from September 1 to November 30 (Doc. 162) and discussed on a phylogenetic classification of the Diptera in general and on more detailed taxonomic questions (see Cumming et al. 2011). He had flewn into the USA on August 13 at Boston (Doc. 163) and spent the last two weeks of August—labelled "Erholungsurlaub" (recreation holiday)—visiting the department of fossils of the Museum of Comparative Zoology at Cambridge (Massachusetts), the U.S. National Museum—Smithsonian Institution at Washington (D.C.), the Department of Entomology of the Natural History Museum of the Illinois State Natural History Survey at Urbana, the Fields Museum at Chicago (Illinois), and the American Museum of Natural History at New York (NY) (Doc. 162). As far as one can judge from his postcards and letters to friends and colleagues, he must have enjoyed this journey. It is quite obvious that during this time he laid the fundament for a faithful and seminal exchange of ideas with North American, especially Canadian, colleagues.

Willi Hennig met colleagues from abroad on a larger scale only at one more occasion outside Germany, when he attended the 12th International Congress of Entomology in Canberra (Australia) in 1974. He used this official occasion for a longer private journey and visited New Guinea, too (Schlee 1978). It seems as if Willi Hennig became fond of long-distance-travels, because he and Irma travelled from February 22 to March 10, 1974, to North India and Nepal, where they visited every touristic place one can imagine (Agra, Khajuraho, Varanasi, Calcutta, Kathmandu, Patan, Bathgaon—they even afforded for a flight to the jungle lodge Tiger Tops). In, 1976, they travelled through South-East Asia, where they saw Hong Kong, the Philippines, Indonesia (Java and Bali), Thailand, and Burma (Doc. 164). These journeys were certainly highlights in Willi Hennig's private life. Except for his stay in North America, he had only seen some European countries, but he had never visited exotic, tropical regions. On a staff questionnaire ("Personalbogen") of 1953, Hennig had answered the question "Have you ever been abroad" with "only as a soldier" (Doc. 103: 12). Since there is a fairly complete record of Hennig's journeys (passport stamps, postcards from vacations, official reports, information provided by his widow and sons), Willi Hennig had certainly not made a longer voyage before 1974 than the return trip from Australia.

The Hennig family lived in an unconspicuous house (Fig. 48) in Ludwigsburg-Pflugfelden. Although Willi Hennig lived there for 13 years, he kept a certain distance to the Suebian environment around (Doc. 196). Through all these years, he did not become familiar with the Suebian

Fig. 48. Ludwigsburg-Pflugfelden, Denkendorfer Strasse 16—the Hennigs lived in the flat on the ground floor (lowest row of windows). Original, taken 29.10.2012.

dialect, he got hardly into contact with Suebian cuisine, and he did not estimate it—he disliked the Suebian specialties, e.g. "Spätzle" (a special handmade type of noodles) or "Flädlessuppe" (fine cut crêpes in a bouillon), and his youngest son Gerd cannot recall evidence that his father enjoyed the Suebian landscape—according to his memories, his father rarely went on excursions alone or with his sons into the Southwest-German region. Mentally, he—and Irma, too—remained residents of Berlin by choice (Fig. 49).

According to Vogel & Xylander (1999), Willi Hennig was for quite a number of days absent from school as a boy, due to health problems. His medical examination in 1936 did not reveal any health risks (Doc. 060), but since he got wounded in 1942, and suffered from jaundice when he was a POW (Doc. 078), one cannot exclude that certain damages to his health remained. At any rate, Willi Hennig had a "light heart attack" shortly before he left for Canada in 1967 and another one immediately before his return flight (Schlee 1978). Irma Hennig had told her sons that this latter heart attack was actually really serious—the Canadian physician had recommended that Hennig should postpone his return trip to Germany (Doc. 201). A medical report of August 1976 confirmed that an additional cardiac infarction must have occurred and had not been noticed (Doc. 165).

Fig. 49. Willi and Irma Hennig in their flat in Ludwigsburg-Pflugfelden, 1971. Courtesy of Irma Hennig (1995).

Although the physician documented that there were neither discomfort nor complains, Willi Hennig's real bodily condition was certainly not as good as one—perhaps including him—could see from outside. After all, he declined to follow an invitation to give an invited lecture to the annual meeting of the German Zoological Society (DZG) in June, 1976, where he could have presented his point of view against Ernst Mayr (who actually came). He gave as reason in a letter to Günther Osche—then president of the DZG—an "enforced rest period" ("erzwungene Ruhepause", Doc. 166). Also on other occasions he gave health problems as an excuse for a denial or a withdrawal.

On November 4, 1976, he had worked on the revision of his "Larvenformen der Dipteren" as usual, and in the evening had enjoyed memories of their last holidays on the isle of Crete. In the night of November 5, he died from a sudden heart attack (Schlee 1978) in his home in Ludwigsburg-Pflugfelden.

His body was buried on the Bergfriedhof in Tübingen on November 10, 1976 (Fig. 50) in a small family circle, his brother Rudolf held the ceremony in a most impressive manner (Schlee 1978), while at the University of Tübingen the flags flew at half mast (Doc. 205).

Fig. 50. Tombstone of Willi and Irma Hennig on the "Bergfriedhof", Tübingen. Original, taken 16.02.2012.

Necrologies have been published by Ax (1977 a, b), Byers (1977), Colless (1977), Kiriakoff (1978), Kühne (1978), Schlee (1977, 1978), and Schuh & Wygodzinsky (1977), a list of his publications appeared in Beiträge zur Entomologie (Anonymus 1978). In "*Darwin & Co.*", I published a biographical book chapter on Willi Hennig in German (Schmitt 2001), additional biographical aspects can be found in Peters (1995), Pont (1981), Schlee (1981), Dupuis (1990), Schmitt (2002, 2003, 2004, 2010, in press) and Vogel & Xylander (1999).

Willi Hennig had received the Fabricius medal of the German Entomological Society in 1954 (see fig. 55), the Gold medal of the Linnean Society of London in 1974, and the Gold medal of the American Museum of Natural History (New York , see fig. 72). He had become a Corresponding Member of the Finnish Entomological Society in 1955, a member of the German Academy of Natural Scientists Leopoldina in 1959, a Foreign Member of the Royal Swedish Academy of Sciences in 1972, a Corresponding Member of the American Entomological Society in 1955 and in 1976 its Honorary Member, and an Honorary Member of the Society of Systematic Zoology in 1976. He had been awarded an Honorary Doctorate (Dr. h.c.) of the Freie Universität Berlin in 1968 (see fig. 46). Especially this last distinction meant a lot for him, as his soul had remained in Berlin (Schlee 1978).

Fig. 51. Willi Hennig on July 30, 1965, sitting on the balcony of his home in Ludwigsburg-Pflugfelden. Courtesy of Gerd Hennig, Tübingen.

CHAPTER THREE

WILLI HENNIG'S PERSONALITY—THE SHY REVOLUTIONISER

All contemporaries who ever met Willi Hennig agree in their description of one certain aspect of his personality: He is described as being cautious, or introverted, or shy (Fig. 52). Dieter Schlee reported (Doc. 117) that Willi Hennig avoided situations where he had to talk to more than four people ("Mehr als vier Leute, und schon war's brenzlig"—"more than four people, and it became immediately ticklish"). Not much had changed since Professor Grimpe's assessment in 1933 (Doc. 024) that Willi Hennig "gives a somewhat shy and unadept impression on first sight". Hull (1988: 132) characterises Hennig as "a very shy and self-effacing man". So he was, and he knew that. In a letter to Lars Brundin (Doc. 169), he confessed that he did not like symposia with many attendants but preferred by far discussions with one or two partners.

On the other hand, he could develop great eagerness and endurance in explaining something to a person, when he gained the impression that his counterpart was indeed seriously interested, as Konrad Senglaub (2001) reported. Senglaub was a student at the University of Leipzig in winter semester of 1946/47. He describes Hennig as a kind and cautious teacher and scientist, void of any academic preciousness. Never had he played up his role as superior, neither towards students nor towards staff. On the contrary, he gave the impression that he felt uncomfortable if he could not avoid to behave as the boss.

He was not a brilliant speaker. The excuses he found in the letters in which he declined to follow an invitation to give a talk appear almost anxious. For example, he wrote to Professor Klaus Sander—who had asked him to give a talk to undergraduate students in quite a relaxed atmosphere in a farmhouse in the Black Forest—that he would like to come but felt he was not in a condition to prepare even the shortest talk (Doc. 170). He also was not very assertive when confronted with a group of people. This can impressively be heard from a tape record of a lecture presented by Klaus Günther to the symposium on methods of phylogenetics, held at Erlangen in 1970 (Doc. 177). Willi Hennig had to chair the discussion following the talk. The end of this discussion turned rather chaotic, and he was not able to finish the session orderly.

Fig. 52. Willi Hennig in the 1950s, taken by Wolfgang Hennig. Courtesy of Wolfgang Hennig, Kranenburg.

However, he could—certainly on extremely rare occasions—also show a strikingly different behaviour, as an episode reported by Gerd von Wahlert demonstrates (Doc. 132). One day in the second half of the 1960s, Willi Hennig arrived at the parking lot of the museum in Ludwigsburg and wanted to place his Peugeot at the usual site, but it was occupied by a lorry. When Hennig asked the lorry driver to remove his vehicle, this man followed very slowly and grumbled "you will get your executive's sleep early enough" ("Sie werden noch früh genug zu Ihrem Beamtenschlaf kommen"). Hennig felt so insulted that he, in a complete flurry, slapped the lorry driver's face. Afterwards, he needed hours to calm down.

Amusingly, those colleagues who knew about the incident kept it private, and they regarded Hennig's behaviour as a sign of human touch.

In contrast to his friend Klaus Günther, Willi Hennig never showed a sign of humour in scientific affairs, neither in letters nor in publications. The terms he coined or the names he gave to newly described taxa are without exception austere in spite of renowned models as, e.g., Carl Linnaeus who assigned the military ranks in his "officers of the flora" ("florae officiarii") according to their bearers' obedience to his ideas. He ranked himself as general, his followers as major general, colonel, etc, but his strong opponent Siegesbeck only as sergeant (Vita III: 155 in Malmeström & Uggla 1957)—and this is only one example of Linnaeus' humerous attitude. A more recent example of a combination of systematics and humour is J.S. Farris' computer programme HENNIG86 (Farris 1988). A certain command of this programme is named "XX", which does not appear very meaningful. Insiders, however, know that "XX, spelled out Spanish as "Dos Equis" is a brand of Mexican beer that has quite some aficionados in the USA, too. The exit command is "yama", which is the name of an Indian death goddess (it "kills" the programme). Under these auspices, one might speculate that Farris chose the misspelled version of Gareth Nelson's name on purpose, when he named a command for buidling consensus trees "nelsen".

However, in a private circle, Willi Hennig could exhibit his cheerful side (Fig. 53). He was highly fond of Klaus Günther's sort of humour (typical for Berliners), he liked to tease colleages and friends by letting them guess where he had spent his holidays, he could play a battle with Gerd von Wahlert in "who can cite more classical authors" (preferably in their languages, and only exceptionally in German), and he could hartily laugh on jokes with the ladies in the departmental administration (Schlee 1978). Also, he exchanged letters full of wit and jokes with Klaus Günther and with other colleagues who were close to him, e.g. his former teacher Willy Matthes and his professor at the University of Leipzig, Arno Wetzel.

If, what Willi Hennig worked out, really meant a revolution to science, the question might stand to reason if Willi Hennig as a person was a revolutionary. Did he intentionally threaten the taxonomic establishment of his days? To consider this possibility I find it useful to follow Frank J. Sulloway's approach of estimating human personality. In 1996, he published his comprehensive analysis of more than 6000 biographies with respect to the factors that make a person a "rebel", i.e. someone who is open to innovations and prone to transcend traditional limits. Sulloway

Fig. 53. Willi and Irma Hennig, Hildegard and Klaus Günther, in Hennig's flat in Berlin-Steglitz, in a merry circle. October, 1955, taken by Wolfgang Hennig. Courtesy of Wolfgang Hennig, Kranenburg.

found that of all factors taken into the meticulous statistical analysis only one explained consistently and significantly the probability of someone to become a "rebel": birth order. His study revealed clearly that laterborns are definitely more receptive to scientific innovations than firstborns, while firstborns tend to be more conforming and traditional.

Willi Hennig was the firstborn of three sons. According to Sulloway (1996), we would not expect him to purposefully revolutionise a branch of science, since for that enterprise "receptiveness for innovations" would be a prerequisite, while a "more conforming and traditional" attitude would be a hindrance. As all contemporaries witness, Hennig was not at all a "rebel" personality. He was unconfident, especially when confronted with an audience of more than four people, he did not write or behave demanding, he did not try to convince someone in personal encounters. Instead, he reiterated what he saw as improvements of systematics in quite a number of taxonomic publications. In letters, he stated that his new method could only be propagated through examples, given by experienced taxonomists. With very few exceptions (1966, 1971, 1974 and the different editions of the "*Taschenbuch der Zoologie*"), he did not address a general scientific readership outside entomology. Obviously, he had planned to publish a textbook of phylogenetic systematics, the introduction of which was published posthumously by Willi Hennig's eldest son Wolfgang (no. 167).

How, then, could it be that Hennig did not end as an extremely specialised—however highly respected—taxonomist of Diptera but became known as the founder of a fundamentally new scientific school? Sulloway's analyses revealed some interesting interactions of birth order and other biographic and social parameters. He found that firstborns of lower social classes were nearly as open to innovations as laterborns of all classes if they were older than 21 when their parents died. This is exactly the case with Willi Hennig. He was 34 when his father died, and 52 when he lost his mother. Thus, this factor could clearly compensate for his status as firstborn.

Sulloway found an additional influence that contributes to the receptiveness to innovations of firstborns: shyness, which interacts in a non-additive manner with birth order. He could demonstrate that the receptiveness of laterborns for scientific innovations was the higher the less shy they were, whereas shy firstborns are as open for innovations as shy laterborns and lose receptiveness for innovations when they lose shyness. Willi Hennig was a firstborn and definitively a shy person, thus appearing as "open for innovations" as shy laterborns. Consequently, it is most probably exactly his shyness and modesty that made Willi Hennig—although a firstborn—a scientific "rebel".

In spite of his "revolutionary" work, he showed downright several traits which are, according to Sulloway, characteristic for firstborns: He was conservative in lifestyle and political attitude and he regarded order as highly important. I believe it is justified to assume that Willi Hennig's quest for a general and reliable phylogenetic system is a consequence of his basic inclination towards ordering the world around him.

Although Hennig was an academic teacher at four different universities (Leipzig, Potsdam, Berlin, Tübingen), he did not himself build an own school of thinking in the academia. Only few of his doctoral students became disseminators of his ideas. Willi Hennig's scientific outreach bases essentially upon his own publications (most important probably 1965, 1966, 1982, 1984, less the early papers 1947, 1949, 1950), and on the commitment of his academic friends and supporters (an incomplete selection from the German speaking area: Ax 1984; Günther 1956, 1962, 1971; Königsmann 1975; Peters & Klausnitzer 1978; Schlee 1971). I find it remarkable that Walter Zimmermann, from whom Hennig took a number of arguments in the *Grundzüge*, did not cite Hennig in 1953, and cited but not mentioned him in 1967, in contrast to Wolfgang Wickler, who followed Hennig's argumentation and used his terminology in his chapter of the same volume as Zimmermann (1967). Wickler discussed the possibility to exploit behavioural traits as characters in a phylogenetic analysis. His conclusion was: yes, as long as we apply Hennig's method and base the phylogenetic hypotheses on synapomorphies exclusively. The preponderance of the traditional approach in systematics can be seen, e.g., in Günther Osche's handbook chapter (1966), in which he cited Hennig several times, mentioned the terms apo- and plesiomorph, but practically ignored Hennig's method and emphasised instead Remane's homology-research.

On the other hand, Willi Hennig was quite selective in accepting others' new thoughts. Although he had inserted a chapter on "evolutionary ecology" in the *Stammesgeschichte der Insekten* (no. 138), he did not take notice of—at least he never cited—the attempts of Gerd von Wahlert to describe phylogeny as an ecological process (1961, 1973).

Hennig had treated the chances of using chemical and immunological data in phylogenetics in the *Grundzuege* already (pp. 162–166), and again in *Phylogenetic Systematics* (pp. 103–107). This demonstrates that he was principally open to new methods and practical approaches. For quite understandable reasons, he did not apply these—at his time—new methods himself, as he had no opportunity to learn them during his student times, and at the institutions he worked existed no laboratories to conduct such studies. When his son Bernd worked as a post-doc at the Max-Planck-Institute for Experimental Medicine in Göttingen, Willi and Irma visited him in 1971. On that occasion, Hennig expressed his desire to understand more about "molecular evolution", and asked his son to teach him on this subject. Regrettably, they could not carry out the intended project, due to complicated health problems in the family and lastly due to Hennig's early death (Doc. 201).

The status of Hennig's Phylogenetic Systematics in biology is nowadays definitively non-controversial (see Dupuis 1979, 1984; Hull 1989). That this is true not only in the German-speaking world (e.g., Peters 1971 a, b; Löther 1972; Ax 1984; Sudhaus & Rehfeld 1992; Jahn 1992; Schmitt & Speck 1992; Lorenzen 1994; Wägele 2000; Wiesemüller et al. 2003; Schmitt & Misof 2009) but also on an international scale, is doubtlessly due to Hennig's English-speaking advocates, especially Lars Brundin (e.g. 1965), Gareth J. Nelson (e.g. 1971), James S. Farris (e.g. 1970, 1988), Arnold G. Kluge (e.g. Kluge & Farris 1969), and Edward O. Wiley (1981). The botanists considered Hennig's approach only with a remarkable delay. Bremer & Wanntorp

(1978) were probably the first to introduce Phylogenetic Systematics to botany. Nevertheless, cladistics became the pradigm for systematics also in this discipline.

An additional important, but in Central Europe widely underestimated, factor promoting the spreading of Hennig's approach worldwide is its substantial accordance with Warren H. Wagner's method of groundplan reconstruction (see Schmitt 2003). I discuss the relation between Hennig's Phylogenetic Systematics and Wagner's Groundplan Divergence Method in chapter 5.5.

Fig. 54. Willi Hennig (standing on the ladder) and Hans Sachtleben in the library of the DEI, 1960. Courtesy of the SDEI.

CHAPTER FOUR

THE TAXONOMIST

4.1. *Taxonomy of Extant Animals* (Fig. 55)

Hennig used the term 'taxonomy' and 'phylogenetic systematics' as synonyms (1950, no. 76: 35ff.), similarly Mayr et al. (1953: 3), who stated that "in modern usage" the terms 'taxonomy' and 'systematics' "are used interchangeably in the fields of plant and animal classifications". I find a distinction sensible and use 'taxonomy' to consider the act of identifying and describing taxa and sorting them into an existing order (Mayr's 'alpha-taxonomy'—1969: 24), while 'systematics' is the act of investigating on and establishing such an order (as defined in Schmitt & Misof 2009: 315f.). This order is traditionally named 'natural system', which is in Hennig's view necessarily a 'strictly phylogenetic system'.

Willi Hennig's start into the world of science—as inferred from his publication record—did not evince that he would become one of the world's leading fly taxonomists. Nevertheless, even his very first scientific paper (Meise & Hennig 1932) on the snake genus *Dendrophis* contains the roots of what later became Willi Hennig's original method. The snakes investigated belong to the genera *Dendrelaphis* (at that time called *Dendrophis*) and *Chrysopelea*. In both genera are snakes that are able to jump, or better to fling or hurl themselves into the air from tree limbs, thereby stretching their body. The *Chrysopelea*-species spread out ribs to both sides of the body, thus gliding over a certain distance. The agamids of the genus *Draco* also produce sort of wings by expanding their ribs. Willi Hennig pounced avidly on the challenge to clarify the difficult taxonomic and nomenclatorial problems, and the papers produced demonstrate that he succeeded surprisingly well for his age. These papers on reptiles are still useful up today. Already then, Willi Hennig learned to cope with nomenclatural problems, with taxonomic descriptions, and with zoogeographic data. The revision of the genus *Draco*, 67 pages long, is an example of careful observation and measuring and show that Willi Hennig mastered quite a lot of statistics (in later papers, he rarely returned to statistical analyses).

Remarkably, the more general problems which came to Hennig's mind when he studied the morphological characters and the nomenclatural

Fig. 55. Fabricius medal, awarded to Willi Hennig in 1953 by the German Entomological Society. Courtesy of Gerd Hennig, Tübingen (photo: M.S., taken 30.10.2012).

history of the treated taxa arose from the geographical distribution of the animals in question. This applies also to Hennig's first paper on flies, a revision of the family Tylidae (now Micropezidae). The "normal" development of a developing taxonomist is to start with a minor faunistic paper ("... a new record of species x from here or there ..."), proceeding over single species or subspecies descriptions ("a new species in the genus XX from here or there ..."), and only after having gained a certain experience he or she would dare to approach a revision of a larger taxon, as a family. Not so Willi Hennig—he published one genus name, one new name for a subgenus, and nine new species in the first of the three issues of part one alone. In the whole revision, he described more than forty new species. The least one can say is that this approach is unusual. The only plausible explanation is that Willi Hennig has been energetically encouraged by Fritz van Emden.

Another remarkable aspect of these early taxonomic papers is the fact that Hennig included identification keys throughout. Who ever tried to elaborate a key for about ten species as his or her first attempt can really judge on Willi Hennig's achievement in his first taxonomic paper: he developed keys for 428 species in the family Micropezidae.

The countless descriptions of new Diptera species are of practical value only for a limited number of specialist workers. Their general scientific importance bases upon the fact that by far most of them are not just isolated species description but that Hennig embedded them in treatises on more comprehensive biogeographic, morphological, or phylogenetic questions. In addition, he drew the morphological characters in stunning detail, as demonstrated by Fig. 56. Accordingly, even his alpha-taxonomic publications (those in which only or mainly new species are described) are outstanding and can serve as models in their realm.

This principle—to treat taxonomic topics not isolated from other, more generalised aspects as biogeography or phylogeny—can be seen already in the early papers on reptiles, jointly written with Wilhelm Meise. Certainly have Hennig's ideas on biogeography—and later on phylogenetics—found acceptance outside entomology only so late and so slowly because he published them generally within an explicitly entomological context. However, there is a noticeable exception from this principle: Hennig's paper (1938) on the larvae of the most important German (sic!) chrysomelines. He argued that the main reason to study these insect was their economic importance. Often entomologists working in the field of plant protection had to identify beetle larvae of this subfamily, as many leaf beetles can be nasty pests. Only few other papers of his contain aspects

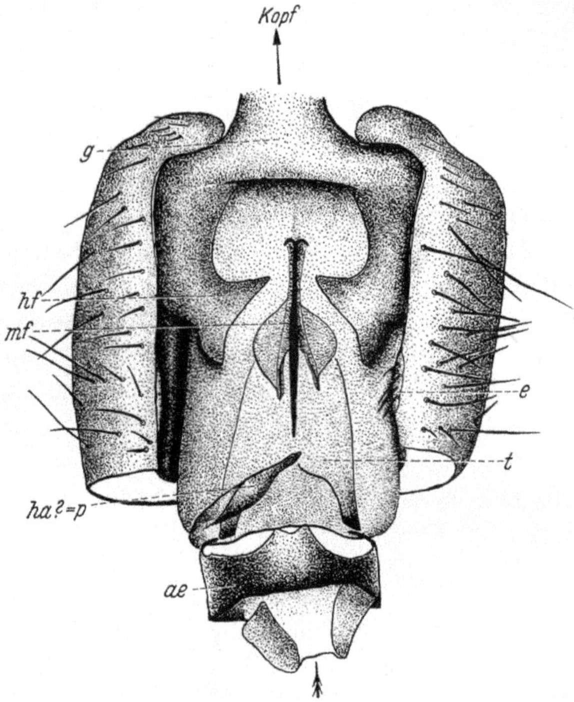

Fig. 56. The copulatory apparatus of *Glossina brevipalpis*, fig. 1 from Hennig's treatise on the male copulatory apparatus of *Glossina* (no. 20).

of applied entomology, to mention especially some publications of the era at the Military Medical Academy on carrot-infesting flies and malaria-transmitting midges. Pest control was definitely not a field of Willi Hennig's ambition. The more we must appreciate that he contributed a 166-pages chapter on the Diptera to the handbook of plant diseases ("Handbuch der Pflanzenkrankheiten") in 1953. Whether he followed his own motivation or the policy of the DEI in the then still young GDR is not known. Jointly with Gerhard Piekarski and B. Sibbing he authored two sections (on the geographic distribution of leishmaniases and sand flies in Central- and South America and in Asia) with six pages of text and 41 maps of the world atlas of epidemic diseases in 1957 and 1959, respectively, presumably because this atlas was edited by Ernst Rodenwaldt, the former director of the Institute of Tropical Medicine where Hennig worked from 1942 to 1945.

Besides these publications, there are only few and very short papers on pure alpha-taxonomy, i.e. identifying and—where necessary—describing new species. Examples of such papers are nos. 90 and 94, which report on Diptera from the Juan Fernandez Islands (Chile), or no. 109 in which Hennig described two new fly species as a contribution to the Festschrift for Erwin Lindner in 1963.

In several papers of the late 1930s and the 1940s, Hennig touched a subject that obviously became the starting point for him to elaborate the method of phylogenetic systematics. It was the fact that in numerous cases the characters of larvae and of adults of holometabolous insects seem to evince different relationships. In 1943, he discussed this matter explicitly, and he treated the topic in detail in his "Larvenformen der Dipteren" (1948–1952). He found the solution of the problem of incongruent distributions of characters by discriminating between those, which have been retained unmodified by the organisms in the course of their evolution, and those, which have been altered. Only by the latter hypotheses of phylogenetic relationship can be substantiated (which is discussed in the following chapter).

The shere number of Hennig's taxonomic papers is overwhelming—nearly one hundred of his over 160 publications can fairly be rated "taxonomic". In many cases, the emphasis on biogeography or on morphology is nearly equal to the taxonomic content. Until 1945, he contributed twelve parts with 431 pages and 26 plates, treating 13 Diptera families, to Erwin Lindner's *Fliegen der paläarktischen Region.* The families covered by these chapters are the Tanypezidae, Coelopidae, Milichiidae, Carnidae, Odiniidae, Braulidae, Otitidae, Ulidiidae, Megamerinidae, Diopsidae, Psilidae, Piophilidae, and the Platystomidae. After WWII, he authored another three parts of "the Lindner", on Sepsidae (91 pages, 10 plates), Muscidae (1110 pages, 33 plates), and Anthomyiidae (974+LXXVIII pages, 114 plates). The scientific importance of these chapters can simply not been overestimated. When regarding the number of pages and illustrations—thousands of them he drew himself—one can really understand that he so often apologised for not following an invitation by pointing to his immense workload. For him, "the Lindner" was certainly an affair of his heart, as one can see from numerous letters in the archive of the SMNS.

The 337-pages chapter of the *Handbuch der Zoologie* (no. 148) is an outstanding example of Willi Hennig's diligence and originality. Many other chapters in the *Handbuch der Zoologie* discuss the available literature and compile their results. In contrast, Willi Hennig analysed data on morphology, ecology, and physiology under a clearly phylogenetic perspective. Again, this is not a pure taxonomic publication.

A last, comprehensive, paper on the classification of the Diptera, has never been published. Willi Hennig had co-operated with the editors of the *Manual of Nearctic Diptera* over years, since his visit in Ontario in 1967 (see Cumming et al. 2011). He had submitted an extensive manuscript, which was, however, only a draft version at the time of his death. Since the editors of the *Manual of Nearctic Diptera* were reluctant to adapt Hennig's manuscript so that it fit with the most recent facts and ideas, they decided not to publish Hennig's contribution. Originally, his chapter should form a whole volume (no. 3) of the *Manual*. Thus, Willi Hennig's dipterological legacy became completed not by a comprehensive paper of his own, but rather by the spirit he implanted in the whole project (Wood 1989).

4.2. *Amber Fossils* (Fig. 57)

Whoever studies phylogeny will sooner or later be confronted with the role of fossils. Willi Hennig mentioned fossils in certain contexts, e.g. when he analysed the wing venation for phylogenetic purposes (no. 84), but he did not investigate on petrifications per se. Since the main objects of his practical phylogenetic studies were delicate flies, it is easy to see why he did not. There are only extremely few conventional fly fossils, at least there were—things have changed remarkably in the meantime, see Bechly (2010). However, as early as in 1936 (no. 16) he wrote that "an interesting possibility for a test of the above described conjectures should be provided by the study of the amber fauna" ("eine sehr interessante Möglichkeit der Prüfung müssten die vorstehend ausgesprochenen Vermutungen durch eine Untersuchung der Bernsteinfauna erfahren"). Moreover, his original plan of a post-doctoral project focused on amber fossils (Doc. 173). He must have kept this idea in mind over years, because in 1938 he wrote a paper on Diptera that had been included in resin in the geological past and become amber fossils that way (no. 31).

In 1964, he got access to the amber collection of the palaeontological institute at the Georg-August-Universität Göttingen. The former director of this institute, Adolf Seilacher, had become full professor at the University of Tübingen that year. He had confirmed that the famous amber fossils from Königsberg (former East Prussia, now Kaliningrad) were kept there (Doc. 178). From then on, Hennig published 17 papers on Diptera of the Baltic amber and three on fossils of the Lebanon amber. In addition, he borrowed material from numerous collections all over the world, and he used the data on amber fossils in his review papers and handbook chapters.

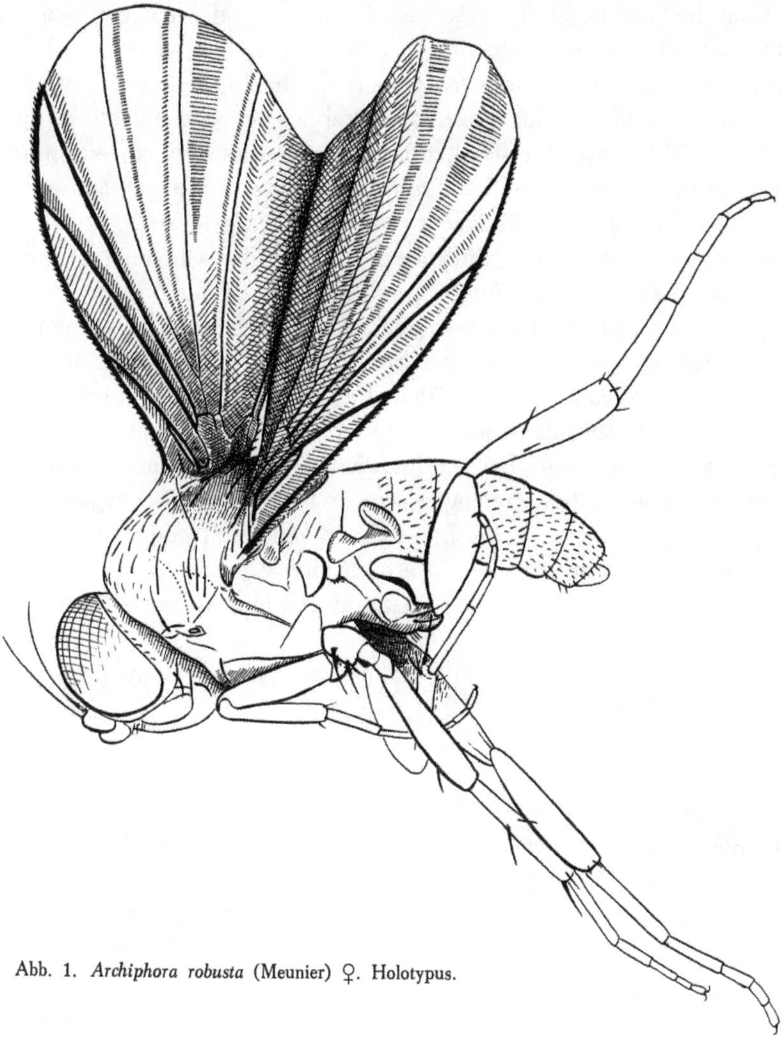

Fig. 57. Holotype of *Archiphora robusta*, a fly of the family Sciadoceridae , embedded in amber, drawing by Hennig (1964, no. 111).

In 1967, Willi Hennig had by chance come across a note on "fossil resin" (which means: amber) from Lebanon (Doc. 179). He could, again by a lucky chance, ask Professor Seilacher, who was going on a geological expedition to Lebanon, to bring amber from there to Ludwigsburg. As he had expected, he found fossil insects in the few hundred grams of amber that Professor Seilacher (or his students) had collected just casually. These were the oldest insect amber fossils known at that time. While the Baltic amber is from the Eocene, i.e. about 40–54 million years old, the Lebanon amber is from the Cretaceous, meaning it is about 130 million years old. Hennig succeeded in receiving funds from the Deutsche Forschungsgemeinschaft so that he could send his assistant, Dieter Schlee, on a collection trip to Lebanon. This marks the starting point of a success story since hundreds of fossil insects were found in amber from Lebanon, which allowed for countless new insights into insect evolution and phylogeny.

Of an especially high significance are the fossils—which means mainly amber-inclusions—in Hennig's *Stammesgeschichte der Insekten* of 1969 (no. 138). In this book, Hennig aimed at dating the temporal sequence of lineage splitting events, which had been found by investigations on extant organisms, by means of fossils. He had exemplified this approach already earlier in some shorter papers on Diptera (e.g. nos. 111, 120, 132). Still in 1969, he held the view that the hierarchical rank of a taxon should be assigned according to its geological age, an idea he had already developed in his *Grundzüge* (no. 76: 216ff.). He pondered to introduce an origin in the Cretacous as the criterion for establishing a "family", while the origin of the "orders" should lie in the Trias and that of the "classes" in the Carboniferous (no. 76: 257).

But besides of these considerations, Willi Hennig emphasised that knowledge of fossils is not a prerequisite for tracing the temporal sequence of species splittings (e.g. in the *Grundzüge* p. 134). This means necessarily that one can principally do phylogenetics without palaeontology. Palaeontologists did, understandably, not appreciate this view. On top of that, Hennig made no attempt to reconcile them. A short but striking example demonstrating the tense relations between Hennig and palaeontologists is an event of October 30, 1953 (Peters 1995). Willi Hennig had been invited to present his new method in a lecture at the natural history museum of Berlin. The following discussion turned somewhat hot tempered, so that Hennig got tense. When the palaeontologist Walter Gross asked him—certainly somewhat idignant—what about the fossils, Hennig replied, also a bit impatient "I am not interested in your fossils" ("Ihre Fossilien interessieren mich nicht"). Whereupon Gross left the audience in anger, shouting "then your theories don't interest me either!" ("dann interessieren mich Ihre Theorien auch nicht") (Doc. 132).

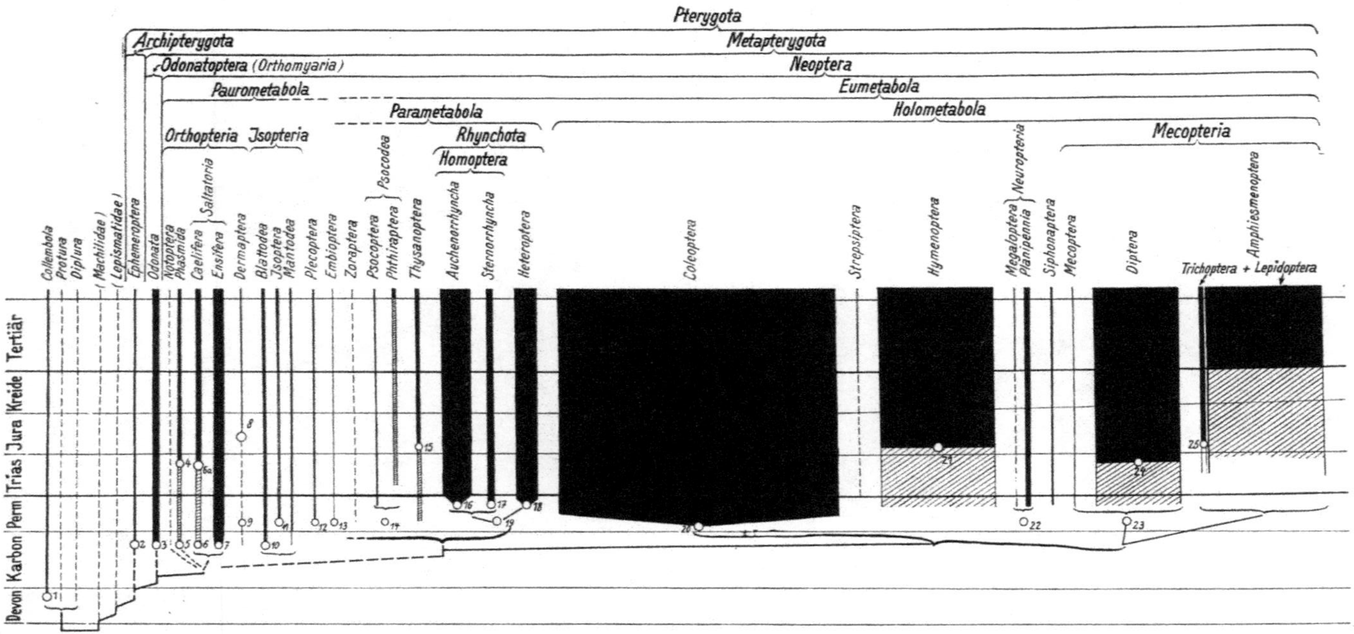

Fig. 58. Phylogenetic tree of the insects, from Hennig (1953, no. 80).

CHAPTER FIVE

THE SYSTEMATIST

5.1. *The Era Until 1960* (Fig. 59)

In the course of the history of biological systematics, the organisms have been ordered according to extremely different criteria, often economic ones: usability, harmfulness etc. Since Charles Darwin's basic work *On the Origin of Species by Means of Natural Selection* (1859) the opinion became generally accepted that a "natural" system should group organisms in a way that reflects their evolution ("descent with modification"). This general consensus did, however, by no means mean a consensus on the question which of the two aspects should be decisive for the grouping, the 'descent' or the 'modification'. If the former aspect is chosen as the criterion, a strictly genealogical notion of "relationships" follows. If the latter is used, "relationship" means formal resemblance or similarity. The first crucial step in the development of Willi Hennig's systematics was an absolute restriction on 'genealogy' when defining 'relationship' (see also Richter & Meier 1994). Hennig made this step quite early, as can be seen in the papers of the 1930s. He wrote in his paper on the geographic variation in the genus *Draco* in 1936 (no. 14: 552): "apparently, both races [of *Draco volans*, from the Philippines and from the Sunda islands] have evolved independently from the Western stem form, consequently they are not immediately related to each other" ("beide Rassen sind aber offenbar unabhängig voneinander aus der westlichen Stammform hervorgegangen, haben also unmittelbar phylogenetisch nichts miteinander zu tun…"). The two forms had been regarded identical before and bore the same taxonomic name, due to external similarity.

In their joint papers, Willi Hennig and Wilhelm Meise followed an entirely conventional, i.e. ambiguous, concept of "systematics" (e.g. no. 1 in 1932). They started with alpha-taxonomy and geographical distribution, they used the term "monophyletic" in the pre-Hennigian sense of "single origin in evolution"; there is no indication of distinguishing between primitive and derived traits. In 1935 (no. 5: 139) they wrote about a "monophyletic derivation" of tooth types and discussed their "phylogenetic and systematic" significance. This is clear evidence that at that time they saw "systematics" not necessarily as "phylogenetics".

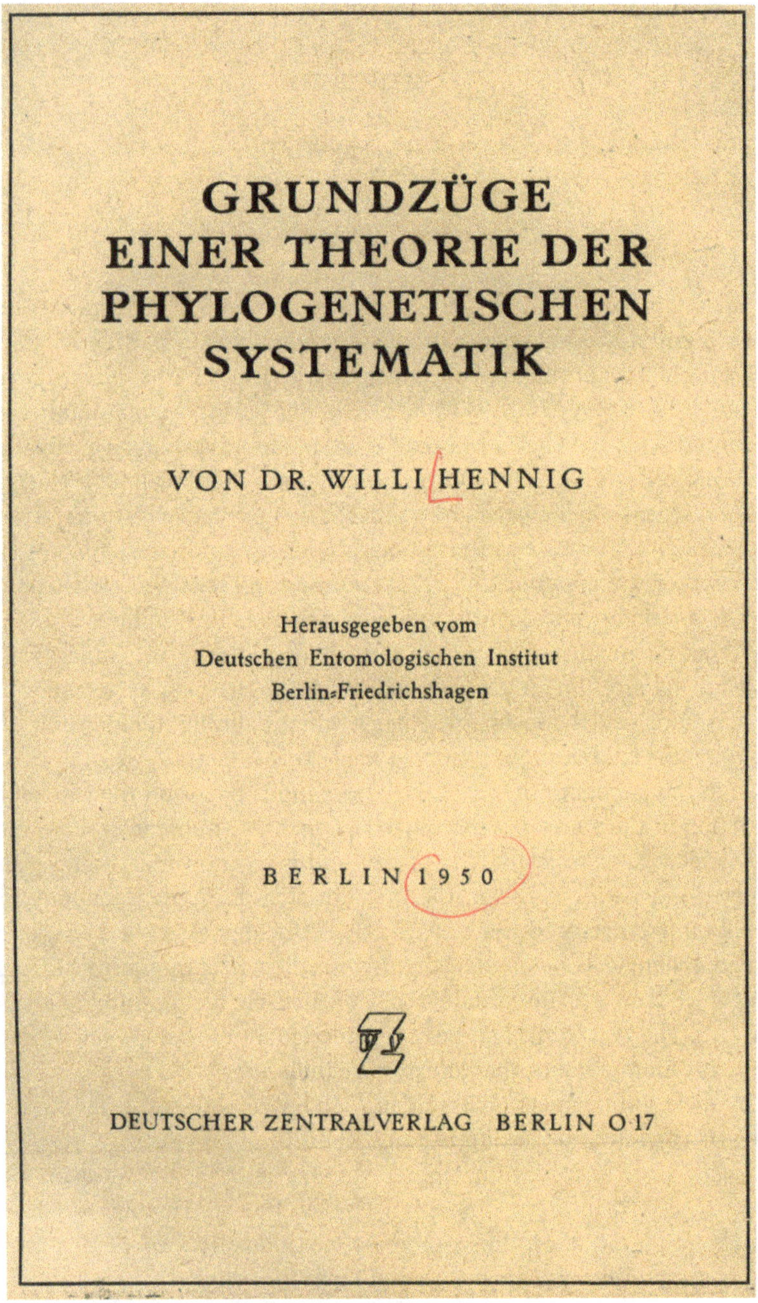

GRUNDZÜGE
EINER THEORIE DER
PHYLOGENETISCHEN
SYSTEMATIK

VON DR. WILLI HENNIG

Herausgegeben vom
Deutschen Entomologischen Institut
Berlin-Friedrichshagen

BERLIN 1950

DEUTSCHER ZENTRALVERLAG BERLIN O 17

Fig. 59. Cover of the *Grundzüge* (Hennig 1950, no. 76).

Box 6 Did Hennig steal from Daniele Rosa?

Since Willi Hennig wrote the manuscript of his fundamental work when he was a prisoner of war, he had hardly any access to scientific literature. He had to rely on his memory and on the letters his wife, Irma Hennig, was allowed to send him. He asked her from time to time to copy—by hand—some passages of papers he needed. Unfortunately, neither he nor his wife kept all of these letters after he returned to Germany. Thus, there is no record of whether or not Willi Hennig took notice of Daniele Rosa's so-called theory of Hologenesis.

According to Arnold Kluge (personal communication 2001), "it has been suggested (....) that key elements of phylogenetic systematics are the same as those in Rosa's (1918) little known publication on the theory of hologenesis, and some have even speculated that Hennig learned of those details during the time he spent in Italy (...). Indeed, it seems certain that Hennig visited Rosa's former student Giuseppe Colosi in Florence, and that he read Rosa's papers at the Institute Library (Baroni-Urbani, 1990: 2). According to B. Lanza (pers. comm.), Hennig's visit to Florence occurred a few years after World War II, but before the 1950 publication of his *Grundzüge*". This description is supported by Gareth J. Nelson (personal communication 1999), with the exception of the dating of Hennig's encounter with G. Colosi.

Rosa had published a book on "La riduzione progressiva della variabilità" (the progressive reduction of variability) in 1899. A German translation had appeared in 1903. Willi Hennig cited this book in 1950 as well as in 1966. In this publication, Rosa treated rules or laws of evolution. Willi Hennig mentioned Rosa when he discussed possible tools of determining the direction of character transformation (character polarisation). He discussed in detail the consequences of the "progressive reduction of variability", emphasised that the phenomenon had been recognised by Gustav Theodor Fechner in 1873 already, and suggested to name it "Fechner-Rosa's law" (no 76: 340). In 1918, Daniele Rosa published the book "Ologenesi—Nuova Teoria dell' Evoluzione e della Distribuzione dei Viventi". A preliminary note had already appeared in 1912, and a summary was published in French in 1923 (see also 1988). A French translation of the book was published in 1931.

In this "theory", Rosa states that species always split into two descendants, and that of these two always one will change evolutionarily at a higher rate than the other. He called the faster changing line "linea precoce" (precocious), the slower changing one "linea tardiva" (tardy). The precocious line should on one hand evolve at a higher rate, but on the other hand keep inferior organisation relative to the tardy line. I could not find a theoretically convincing substantiation for this statement. My impression is that Rosa stated these relations axiomaticly. It is certainly important that Rosa did not provide any empirical criterion as to how to distinguish between the precocious and the tardy line, nor did he even intimate that his "theory" would offer a practical tool for systematics (at least not in the 1923 paper).

According to the 1923 summary, there are two points in Rosa's ideas superficially resembling aspects of Hennig's method: (1) the 'law of ramification' that causes species obligatorily to split into two daughter species, and the normally dichotomous design of Hennigian cladograms. However, Hennig discussed in detail the mode of speciation without starting from dichotomous splitting. On the contrary, in his 1966 book he states explicitely that "if phylogenetic systematics starts out from a dichotomous differentiation of the phylogenetic tree, this is primarily no more than a methodological principle" (no. 125: 210). (2) Rosa's distinction of a "linea precoce" and a "linea tardiva", and Hennig's conceptions of apomorph and plesiomorph. However, Hennig referred not at all to a 'law of different evolutionary speed'. From p. 88 in his 1966 book it becomes clear that he regarded character transformation as a prerequisite for systematists in order to distinguish different species: if we can recognise two (or more) species where before was only one, then at least one character must have changed in each additional lineage. There is no idea of an obligatorily faster evolving lineage whatsoever.

All of his pre-1950 publications demonstrate that Willi Hennig was primarily a taxonomist who wanted to reach a theoretically better substantiated order than the previous ones. His interests in evolutionary processes were clearly motivated by his desire to give good reasons for his hypotheses on character polarisation. Moreover, he wanted to polarise characters in order to establish monophyletic (= holophyletic) taxa. Therefore, it is unlikely that he came across Rosa's book in the course of his literature search. Since Willi Hennig, from his 1950 book on, conscientiously cited other sources that contributed to his method (e.g. Adolf Naef's papers and the seminal paper of Zimmermann 1937), it is pure speculation to insinuate that he had knowledge of Rosa's 'theory' but dismissed it. Thus, even if he had known Rosa's "theory", he would possibly—and understandably—not have seen the need to cite it.

Stunningly, Bernhard Rensch cited Rosa's book in 1947, and Walter Zimmermann cited it in 1953. The latter listed Rosa in the references and mentioned him when discussing internal causes of evolution, but not when he introduced his own concept of "Hologenie". I assume that none of them actually read Rosa's book but relied on the summaries of it. By the way, George Gaylord Simpson cited the French translation in his 1944 book on "Tempo and Mode of Evolution".

After all, Willi Hennig wrote the core parts of his Grundzüge in 1945, during his time as a POW. As shown in chapter 5.1., he had clear conceptions of the basics of his method much earlier than that. The manuscript was completed in 1948 or even 1947 (see no. 58). Thus, a visit of Florence 'after 1945 but before 1950' is irrelevant to the question of what he possibly learned from reading Rosa's book. In addition, it is extremely unlikely that Hennig travelled to Italy within that time frame, as he stated officially in 1953 that he was abroad before only as a soldier (Doc. 103: 12), and as his son Wolfgang confirmed (Doc. 172).

However, here lay the beginnings of the first major contributions Willi Hennig made to systematics: to define "relationship" in a strictly genealogic way, i.e. distinguishing relationship from similarity. However, he did not invent this distinction, but rather took it from the writings of Adolf Naef, especially from his papers published in 1917 and 1919. Naef distinguished explicitly (1917: 40) between "blood relationship" ("Blutsverwandtschaft") and "form relationship" ("Formverwandtschaft"). On the following page, he regretted that the discipline of phylogenetics had to date no sound methodical foundation. In 1919, Naef introduced the term 'Stammart' (stem species) for the species from which any two or more extant species arose in the past, and he mentioned in a casual manner (p. 46) that "presently one performs generally 'phylogenetic systematics' which means that the system should in its hierarchy of classes, orders, families etc. express the phylogenetic relationships, like the phylogenetic tree". Hennig knew (and cited) these papers. That means, he neither invented all these terms nor did he start from Zero when he began to depart from conventional systematics.

It is not clear who stimulated Willi Hennig's development of phylogenetic systematics. Either Wilhelm Meise (a very modest person who did not claim to be the one), or Fritz van Emden, or Klaus Günther must be considered. It is possible that Hennig spoke to none of them nor to anybody else about his scientific ideas. However, this is not very probable, and if we look at the extensive and detailed correspondence between Hennig and Günther after 1963 (Schmitt 1996), it was often Willi Hennig asking for advice. Therefore, it is likely that Klaus Günther played a certain part in the considerable step Hennig made in 1936 (no. 16), when he wrote on the relations between geographic distribution and systematic classification of some dipteran families as a contribution to the problem of classifying taxa of higher order.

The concept of a phylogenetic systematics existed definitely before Willi Hennig outlined his method (see e.g. Naef 1919; Danser 1942). These early phylogenetic systematists knew and accepted, of course, that only homologies can indicate phylogenetic relationship, while adaptational similarities (convergencies, analogies) evolved independently and only dissemble phylogenetic relationship. However, as Naef lamented, there existed neither generally accepted methods for ascertaining homology, nor a scientific method to reconstruct phylogenetic relationships.

Obviously, Hennig had recognised in 1936 already that distinguishing between homologous and non-homologous characters is not sufficient for a precise reconstruction of phylogeny. In his discussion of the relationships between two subfamilies of flies, he argued (no. 16: 170): "Both groups resemble each other in the most characters. Even if we could think about the primitive characters ... that they mean little for a closer relationship as they are simply independently retained inheritage from the ancestor of all tylids, then the progressive characters in common suggest especially the idea of a closer relationship of tylines and neriines" ("Beide Gruppen stimmen in den meisten Merkmalen überein. Könnte man auch hinsichtlich der primitiven Merkmale (...) die Meinung vertreten, dass diese Merkmale als einfach unabhängig bewahrtes Erbe von den allen Tyliden gemeinsamen Vorfahren wenig für die Feststellung einer näheren Verwandtschaft zu bedeuten hätten, so legen die gemeinsamen fortschrittlichen Merkmale (...) den Gedanken an nähere Verwandtschaft der Tylinen und Neriinen besonders nahe").

In this publication we find also a fundamental argument in favour of a primacy of the phylogenetic system over all other systems (p. 172): "Just this case in particular demonstrates clearly that accepting the fact, that the systematic ordering must be done according to a thorough analysis of the phylogenetic relationships rather than according to the degree of morphological differentiation, does not put the work of systematists on uncertain ground:... systems based on phylogenetic relationships will dither, too. However, they will—as long as we believe in the possibility of progress in phylogenetic knowledge at all—gradually lead to a stable final state" ("Gerade dieser Fall zeigt auch besonders deutlich, dass die Anerkennung der Tatsache, dass systematische Einteilung nach der sorgfältigen Untersuchung der phylogenetischen Verwandtschaft und nicht nach dem Grade der morphologischen Differenzierung zu geschehen hat, die systematische Arbeit nicht auf eine unsichere Basis stellt:... auf der phylogenetischen Verwandtschaft aufgebaute Einteilungen [werden] zwar auch schwanken, sich aber doch, wenn man überhaupt an die Möglichkeit eines Fortschritts der phylogenetischen Erkenntnis glaubt, allmählich einem stabilen Endzustande nähern"). This quotation shows, in addition to anything else, also the complicated style of Hennig's prose already in the early phase of his career as a scientific author.

Willi Hennig expresses his point of view quite clearly in the extensive discussion of the problem of the relation between larval and imaginal systematics in 1943 (no. 63: 143): "The system of the animals is in my view quasi a theory about the phylogenetic relationships among the animal forms" ("Das System der Tiere ist nach meiner Auffassung gleichsam eine Theorie über die phylogenetischen Beziehungen der Tierformen"). Through this step the aim of a scientific systematics of the organisms (in this case: the animals) is unambiguously stated: Ascertaining the phylogenetic (= genealogical) relationships. However, still there is no method provided by which this target could be reached. Hennig's first considerations sound rather cautious: "Since the phylogenetic relationship can (principally) not (however, in many cases yet indeed) be seen from the degree of (morphological) similarity between two or more forms or groups of forms, it is important to e v a l u a t e the similarity relations. The task of systematics is not (quantitative) counts and measures of the similarity relations between the organisms, but their (qualitative) evaluation and the distinction between those similarity relations which indicate (independently of their quasi mathematical extension) closer phylogenetic relationships and those indicating less close or no phylogenetic relationships at all" ("Da [die phylogenetische Verwandtschaft] . . . nun (grundsätzlich) nicht (wenn auch in vielen Fällen tatsächlich doch) aus dem Grade der (morphologischen) Ähnlichkeit zweier Formen oder Formengruppen zu ersehen ist, so kommt es also darauf an, eine W e r t u n g der Ähnlichkeitsbeziehungen durchzuführen. Nicht (quantitative) Zählung und Messung der Ähnlichkeitsbeziehungen zwischen den Organismen, sondern deren (qualitative) Wertung und Unterscheidung zwischen solchen Ähnlichkeitsbeziehungen, die (unabhängig von ihrer sozusagen mathematischen Größe) nähere, solchen, die weitere und solchen, die überhaupt keine phylogenetischen Verwandtschaftsbeziehungen anzeigen, ist Aufgabe der Systematik") (p. 140, emphasis in the original).

That Willi Hennig dismisses counting and measuring in the previous paragraph does not mean that he did not exactly count and measure himself. Several of his papers, especially the early ones, witness his aptitude to do exact empirical research. A downright model of such a study is his revision of the genus *Draco* in 1936 (no. 9).

Obviously, the conceptual basis of the method of phylogenetic systematics matured in Willi Hennig's mind during the war times. In his notebook (Doc. 070) we find hand sketches of provisional cladograms (Figs. 60–62) documenting that he pondered on phylogenetic problems during his spare times in Italy. When he wrote the manuscript of his *Grundzüge* during the time in British captivity, he had already a clear concept of its structure and content (see Fig. 31). He introduced the terms "plesiomorph" for independently retained characters (which do not prove closer relationship) and "apomorph" for the "progressive", uniquely derived, characters (which are crucial for the reconstruction of the phylogeny) in a little known publication in 1949 (no. 73). There, he refers to the manuscript of the *Grundzüge*, which he had submitted to the publishers already. On the other hand, he stated on p. 106 and in a footnote on p. 144 in the *Grundzüge* that he had introduced these terms "elsewhere" ("an anderer Stelle"). Analysis of his earlier publications reveal that he could only have

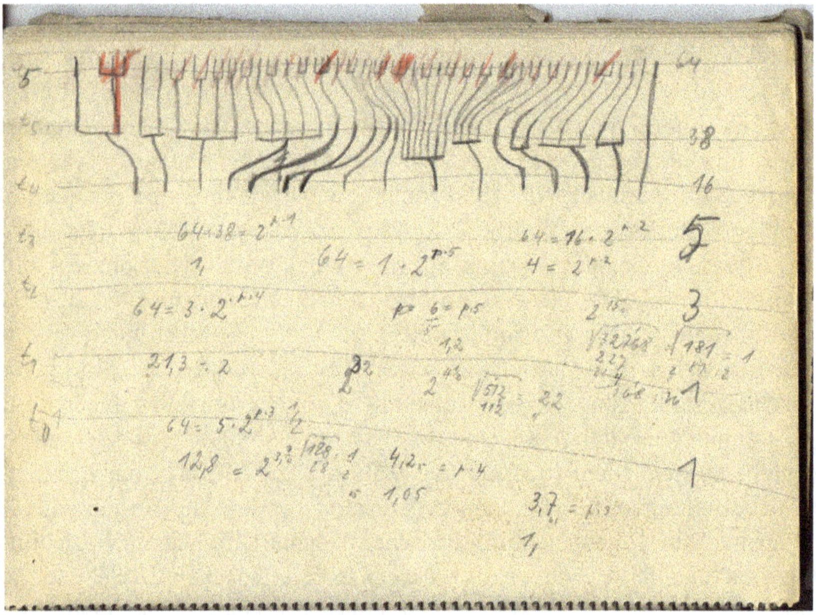

Fig. 60. Sketch of a cladogram in Hennig's notebook of 1944. Obviously, Hennig aimed at finding a numerical regularity in the increase of the number of species over time. Courtesy of Gerd Hennig, Tübingen.

meant the paper of 1949 (no. 73). The only plausible explanation is that he added the footnote in print. This interpretation is corroborated by the fact that Hennig inserted a citation of his 1949 paper on p. 33 of the *Grundzüge* but did not give the bibliography under References.

In addition to the terms "apomorph" (from Greek "apo" meaning "ab-" or "away from" and "morphe" meaning "shape" or "form") and "plesiomorph" (from "plesios" meaning "neighbouring"), he coined a number of terms which have not been accepted by the scientific community, as "apoök"—"plesioök" for the distinction of derived and primitive modes of life, "apochor"—"plesiochor" for derived and primitive geographical distributions, or "stenomer" for groups comprising only few species. When Hennig explained the meaning of 'apomorph' (on p. 111), he presented a schematic diagram (Fig. 63) and described in length the so-called 'rule of deviation', but did not mention the term 'apomorph'.

The generally accepted terms "apomorph" and "plesiomorph" referred originally to groups of organisms (taxa), not to characters (Richter & Meier 1994). Only later made Hennig clear how to use these terms, and explicitly only in a posthumously published little book on *Aufgaben und*

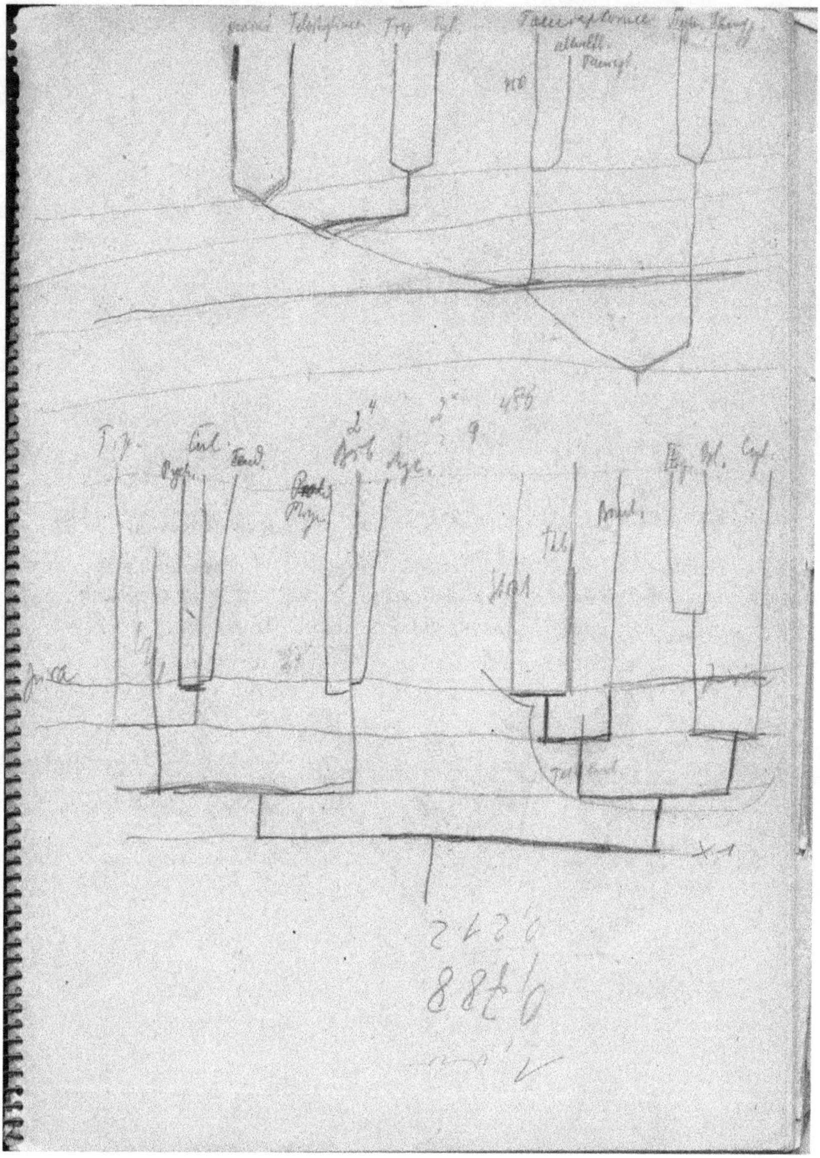

Fig. 61. Sketch of a cladogram in Hennig's notebook of 1944. Here, Hennig mulled over different hypotheses on the phylogeny of Diptera. Courtesy of Gerd Hennig, Tübingen.

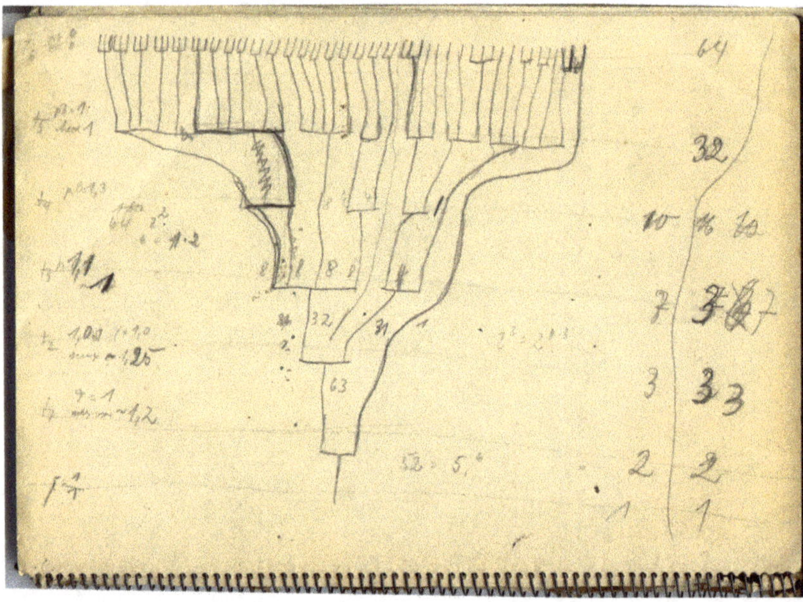

Fig. 62. Sketch of a cladogram in Hennig's notebook of 1944. The figures on this sketch correspond to Hennig's calculations in chapter IIIB of the *Grundzüge*, from p. 313 on. Courtesy of Gerd Hennig, Tübingen.

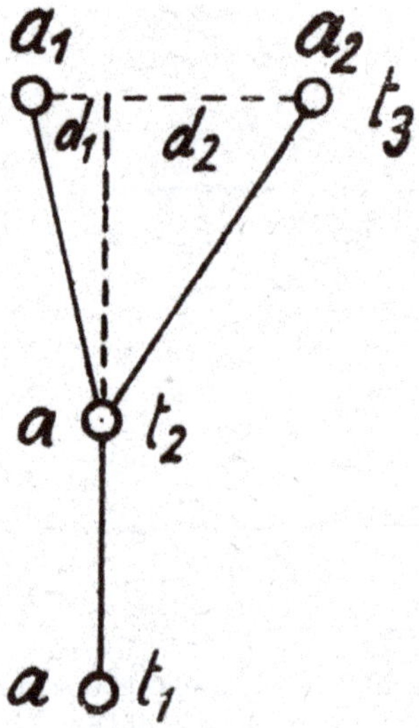

Fig. 63. Scheme illustrating the "deviation rule" (species a exists from t_1 to t_2, splits at t_2 into the plesiomorph species a_1 and the apomorph species a_2 and thereby ceases to exist, or survives in both daughter species), from Hennig (1950, no. 76).

Probleme stammesgeschichtlicher Forschung (1984, no. 167), where he wrote on p. 41 that it is "basically wrong to speak of plesiomorph and apomorph taxa" ("grundsätzlich falsch ... von plesiomorphen und apomorphen Gruppen ... zu sprechen"). Nevertheless applied Hennig these terms already in the *Grundzüge* to characters, which is the only theoretically correct way, e.g. on p. 142, where we read that every species derived from a certain ancestor possesses a mosaic of primitive characters and those derived to different degrees. He uses the term "Spezialisationskreuzung" (specialisation crossing) to explain such mosaics. He must have seen such mosaics of primitive and derived characters in 1936 already (no. 16). It is completely clear in the paper of 1949 (no. 73), where he writes (p. 138) "In many cases we will demonstrate that a taxon is plesiomorph with respect to one trait or, e.g., in one stage of the metamorphosis, but apomorph in other respects so that a general evaluation will be impossible (specialisation crossing). The suggested terms can be used also in these cases, but then just for single traits" ("Vielfach wird eine Gruppe sich hinsichtlich der einen Eigenschaft, oder z.B. in dem einen Metamorphosestadium, als plesiomorph, in anderen als apomorph erweisen, ohne dass eine Gesamtbeurteilung möglich wäre ("Spezialisationskreuzungen"). Auch in diesen Fällen bleiben die vorgeschlagenen Bezeichnungen, dann eben für die Einzelzüge der Gestalt, verwendbar"). In methodological papers published immediately after 1950 (no. 80 in 1953, no. 89 in 1955), he used the terms "apo-" and "plesiomorph" without any comment as attributes of characters.

It is clear from the proportion of pages and illustrations devoted to different fields of phylogenetic research, that biogeography was most stimulating and rewarding for Willi Hennig. On dozens of pages he treated "chorological relations", the "chorological method", the "biogeographical method—investigation of types of vicariance" and the like. Of the 58 illustrations in the *Grundzüge*, 24 show distributions of organisms by means of concrete maps. In contrast, Hennig presented only five general schemes of genealogical relationships, and only four concrete phylogenetic trees of animal taxa.

Hennig introduced in the *Grundzüge* (no. 76, p. 40) the term 'tokogenetic' to denote the relations between the individuals within one species. As he had realised that a systematist, especially one working on preserved dead organisms, can always investigate on individuals only in a certain stage of their life cycle—quite understandable for someone studying holometabolous insects—he used the term 'semaphoront' (p. 9) to denote an individual at a theoretically infinitesimally short span of its life. Thus, scientists do not study, strictly speaking, 'individuals' but 'semaphoronts'. The different 'semaphoronts' of one individual are connected by 'ontogenetic' relations. 'Ontogenetic' and 'tokogenetic' relations are two forms of 'autogenetic' relations. The latter exist within species. The relations between species (and groups of species) are the 'phylogenetic' relations, which to reveal is the aim of systematics.

The 'rule of deviation' ('Deviationsregel') means that "of the two species evolving from one stem species often one of the two daughter species remains more similar to the stem species in its Gestalt-characters than the other, which deviates from the stem species" ("daß von den zwei Arten, die aus einer gemeinsamen Stammart hervorgehen, häufig eine der beiden Tochterarten in ihren Gestaltmerkmalen der gemeinsamen Stammart ähnlicher bleibt als die andere, die sich gestaltlich von ihr fortentwickelt"; no. 76: 106). As already mentioned, Hennig coined the term 'plesiomorph' for the first and 'apomorph' for the latter species. Because the stem species is connected to both daughter species by exactly the same type of relation, it would not be justified to say that it survived in only one of its descendants. Consequently, Hennig suggested that a species ceases to exist the very moment it splits (pp. 102, 111, 142), i.e. when there are no longer tokogenetic relations between the two groups of individuals (= species). The stem species is said to "survive" likewise in both daugther species. This method of delimiting species in time bears some consequences for the units that can and should be named, as Griffiths (1974: 116ff.) discusses extensively. This idea of an obligatory extinction of a stem species through splitting has caused major objections. It was possibly the main reason for the reluctant acceptance of 'Hennig's principle' in the early years after the publication of the *Grundzüge*.

A central claim of Phylogenetic Systematics is found in Hennig's early publications only implicitly (1936 no. 16: 168 and fig. 3), and only later (1947 no. 68: 278) in explicit phrases: In the supraspecific group categories "the species which can nowadays be recognised are grouped according to their origin by the decay of earlier stem species in such a manner that the respective higher categories comprise several lower groups of species of which we can assume that they originated from the decay of one stem species of which no other extant species can be derived" ("sind die in der heutigen Organismenwelt unterscheidbaren Arten nach Maßgabe ihrer Entstehung durch den Zerfall älterer Stammarten derart zusammengefasst, dass die jeweils höheren Kategorien mehrere niedere Artengruppen umfassen, von denen angenommen wird, dass sie durch den Zerfall einer Stammart entstanden sind, von der außer ihnen keine anderen lebenden Arten abzuleiten sind"). Also in the *Grundzüge* we find this claim in different wordings, so on p. 146: "for the phylogenetic systematics, all those organisms belong to a certain taxon of higher level which descended from a common stem species, but on the other hand also only these" ("Für die phylogenetische Systematik gehören dagegen in eine bestimmte Gruppe höherer Ordnung ... alle, andererseits aber auch nur diejenigen Organismen, die von einer nur ihnen gemeinsamen Stammart abgeleitet werden können"), or on p. 276: "The result of the taxonomic work is a hierarchical system of groups the content of which is clearly determined by the fact that they on one hand can only comprise those species which descended from a common stem species, but on the other hand they must comprise all extant species which descended from this stem species" ("Das Ergebnis der taxonomischen Arbeit ist ein hierarchisches System von Gruppen, die ihrem Inhalt nach völlig eindeutig dadurch bestimmt sind, dass sie einerseits nur solche Arten umfassen dürfen, die von einer gemeinsamen Stammart herzuleiten sind, andererseits aber auch all lebenden Arten umfassen müssen, die von dieser Stammart abstammen"). Finally on p. 307 Hennig refers to Ernst Haeckel's term "monophyletic". Haeckel had used that term in his *Generelle Morphologie der Organismen* (1866) in the famous plate 1 ("Monophyletischer Stammbaum der Organismen") and in a footnote on p. 417 in volume 2, but not defined it explicitly. He used it as an obviously self-explaining term. Possibly one could invoke his definition of "phylum" or "phylon" on p. 205 in volume 1 and on p. XIX in volume 2 as "the entirety of all still existing or already extinct organisms which derive their origin from one and the same common stem form" ("die Gesammtheit aller jetzt noch existirenden oder bereits ausgestorbenen Organismen, welche sich von einer und derselben gemeinsamen Stammform ihre Herkunft ableiten"). The question of how Haeckel defined "monophyletic" matters because in the 1970s a lively discussion broke out—mainly led by Ernst Mayr—on the actual meaning of that term (see chapter 5.3.). At this point I emphasise already that neither in Haeckel's nor in Hennig's definition the stem species is a member of the monophyletic taxon that comprises (all of) its descendants.

Hennig stated in the *Grundzüge* (p. 307) "accordingly, only such groups of species can be named as monophyleticly originated which lastly can be ascribed to a common stem species. In phylogenetic systematics only those group formations are justified which can be characterised as monophyletic in this meaning" ("als monophyletisch entstanden können demnach nur Artengruppen ... bezeichnet werden, die letzten Endes auf eine gemeinsame Stammart zurückgeführt werden können. In einer phylogenetischen Systematik haben daher nur Gruppenbildungen überhaupt

Berechtigung, die in diesem Sinne als monophyletisch zu bezeichnen sind"). Rather casually he introduced the crucial addition which later led to the above mentioned most vehement discussion in the history of systematics: "we have to add to this definition—as very extensively substantiated above—that a monophyletic group must not only comprise species which can be derived from a common stem species but, moreover, that it must comprise all species which descended from this stem species" ("Zu dieser Begriffsbestimmung muss nur, wie oben sehr ausführlich begründet wurde, hinzugefügt werden, dass eine monophyletische Gruppe nicht nur Arten umfassen darf, die von einer gemeinsamen Stammart abzuleiten sind, dass sie vielmehr darüber hinaus auch alle Arten umfassen muss, die von dieser Stammart herkommen") (p. 308).

In the *Grundzüge*, it is mentioned just once (on p. 307) that a strictly phylogenetic system must only contain monophyletic taxa, and that these taxa all have to be characterised by the common possession of uniquely derived characters. However, the methods by which such characters or character states can be recognised, are described only vaguely. On pp. 172–178 Hennig provides four "rules for the evaluation of single morphological characters" ("Regeln für die Wertung morphologischer Einzelmerkmale"), which can serve as "indicators of the degree of phylogenetic relationship" ("Indikatoren des phylogenetischen Verwandtschaftsgrades"): 1. frequency of occurrence (characters which occur in a wider circle of species indicate more distant relationships than those which are only common to a smaller number of species, p. 172), 2. ontogenetic character precedence (the temporal sequence in which different characters develop in the course of individual ontogenies is often correlated with the degree of relationship, p. 174), 3. complexity of characters (correspondence in more complex characters is more significant than that in simpler characters, p. 175), and 4. corresponding special characters (resemblances which can be interpreted as chance conformities are more significant than those which are in a tight relation with the mode of life of their bearers or which are, according to experience, subject to certain evolutionary trends, p. 178). This all sounds more after Remane's homology-"criteria" (1952: 28ff.) than like actual instruction for the assessment of (syn-) apomorphies. In fact, Hennig mentioned 'homology' only once in the *Grundzüge* (on p. 176), in the context of 'homoiology' (a term introduced by Ludwig Plate in 1922: 7 and nearly synonymous to 'parallelism'). It could, indeed, be irritating that a phylogenetic method, outlined by a German author, ignores completely a concept that was (and is) commonly regarded basic to all phylogenetic thinking, especially by German-speaking zoologists.

When Willi Hennig thought of and wrote about the splitting of a species, thus becoming stem species of a number of daughter species, he started from the idea that there are just two descendants. He took into consideration that a species could split into more than just two daughter lines, as can be seen from the sketch drawings in his 1944 note book (Figs. 30, 60–62), but a dichotomous splitting is the simplest case. This was important for Hennig not the least because he calculated intensively on the increase of species numbers in the course of phylogenesis. Obviously, he hoped to find a law-like relation between time and number of species. When he discussed the numerical change of the number of species in taxa over time (pp. 313–325), he used normal growth curves and found it plausible to assume that there must be an upper limit of species numbers. Thus, he preferred a sigmoid curve like the ones used in ecology to demonstrate population growth up to the capacity limits of a given biotope. Similarly, Hennig argued that the number of species could increase only to limits provided by the resources. He suspected that in younger taxa the speciation rate is higher and will decrease in the course of time. However, he also discussed the possibility that certain "extreme" subtaxa keep their high speciation rate on the cost of others, which will produce species at a lower rate or finally go extinct (Fig. 64). In both cases, a maximal number of species will result, determined by the resources of the environment. Nowhere in this chapter did he mention a reason why species should split dichotomously.

In the *Grundzüge*, Hennig invoked ecological and genetic models to a surprisingly high degree, compared to his later publications. He treated the "problem of the dichotomy of the phylogenetic tree" ("Problem der Dichotomie des Stammbaumes") on two pages (332f.) only. His first argument is that "it is a well known fact that many animal groups of the system each consist on the next lower level of order of two coordinated subtaxa" ("Es ist eine bekannte Tatsache, daß viele Tiergruppen im System in je zwei koordinierte Teilgruppen der nächstniedrigeren Ordnungsstufe 'zerfallen'", p. 332). Subsequently, he considers several reasons why a dichotomously structured tree represents correctly the true phylogeny with a certain probability. He refers to his fig. 54 (Fig. 64) and infers that under the given conditions after only four steps of splitting all species can possibly be arranged in a strictly dichotomous hierarchical system. Although he emphasises that his considerations demonstrate only possibilities, he doubtlessly regarded them as a justification of his normally dichotomous splitting schemes. I come back to this point in chapter 6.

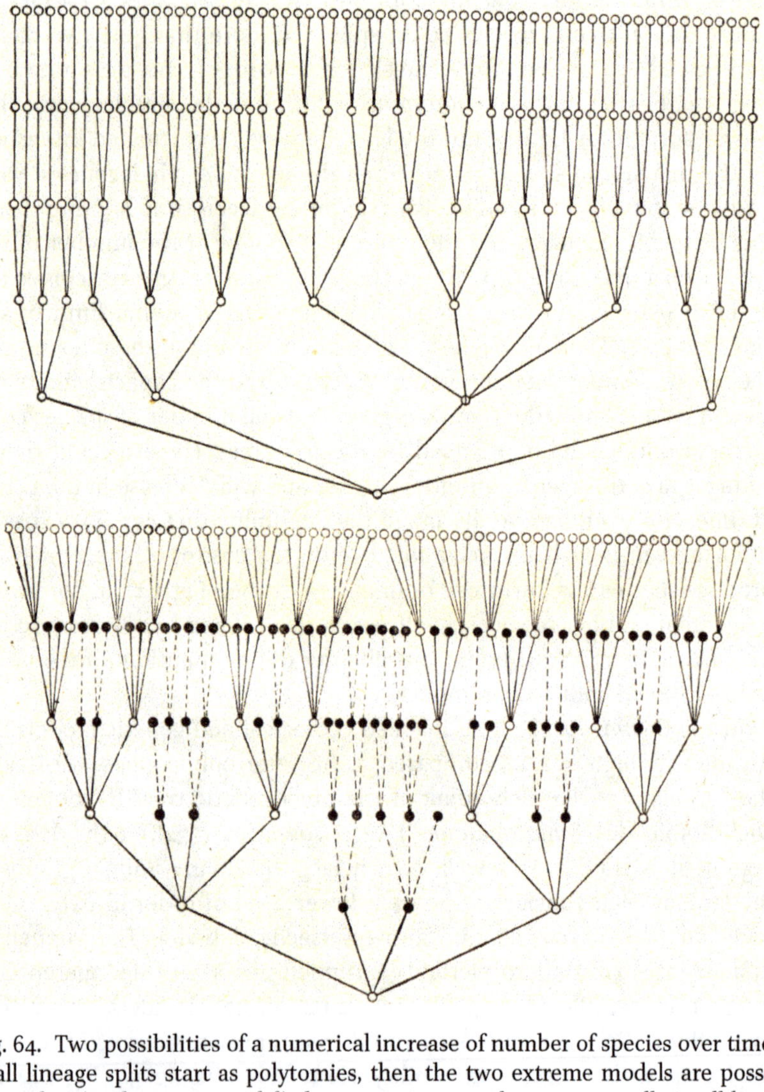

Fig. 64. Two possibilities of a numerical increase of number of species over time. If all lineage splits start as polytomies, then the two extreme models are possible: either (in the upper model) the speciation rate decreases equally in all lines towards the capacity limits of the space, yielding a polytomous tree, or (in the lower model) only the "extreme" daugther lineages keep their rate of speciation, at the cost of the remaining lineages, thus finally yielding a dichotomous tree (from Hennig 1950, no. 76, p. 321).

Only in 1953, in "critical remarks on the system of the insects" (no. 80) Hennig made clear that apomorph characters and their plesiomorph alternatives are homologous. This paper is—in contrast to the *Grundzüge*—written in an extrordinarily clear and intelligible style. Here, Hennig discussed Adolf Remane's homology concept and criticised its lacking distinction between "criterion" and "definition" (p. 11). Remane had explicitly denied that 'homology' could be defined in a conventional manner (1952: 30). Instead, he pleaded for 'identification' of homologies by the application of several 'criteria' (which are, strictly thinking, not logical criteria but empirical indices). Hennig, however, claimed "that as many and as good criteria cannot replace a definition, but that rather any criterion gets its role only through its relation to a definition" ("dass noch so viele und noch so gute Kriterien eine Definition nicht ersetzen können, dass vielmehr jedes Kriterium erst durch seine Beziehung zur Definition zu einem solchen wird").

In this paper (no. 80), Hennig introduced the distinction of 'synapomorphy' and 'autapomorphy'. The former means correspondence of several taxa in relatively derived characters, while the latter denotes a relatively derived character that occurs only in one (terminal) taxon and can, consequently, not indicate phylogenetic relationship of several taxa. Moreover, he discusses (on p. 16f.) the relation between 'convergence' and 'synapomorphy': characters appearing synapomorph can, of course, have become independently similar, so that in reality their correspondence bases on convergence. He argued that "even a little thought shows easily that the phylogenetic systematics would loose any ground below its feet if it would regard all synapomorphies being convergencies in the first place and then demand the proof to the contrary in each case. Rather the burden of proof has to be put on the claim that certain synapomorphies can only base on convergence" ("Ein wenig Überlegung zeigt aber leicht, dass die phylogenetische Systematik allen Boden unter den Füßen verlieren würde, wenn sie alle Synapomorphien zunächst als Konvergenzen auffassen und in jedem Falle den Beweis des Gegenteils verlangen wollte. Die Last des Beweises muss vielmehr der Behauptung auferlegt werden, dass bestimmte Synapomorphien nur auf Konvergenz beruhen können"). I doubt that this concept is theoretically sound—the procedure of starting from convergence and subsequently substantiating synapomorphy is more laborious than the other way round, but the result would be the same. Anyway, most phylogeneticists probably share Hennig's view in practice.

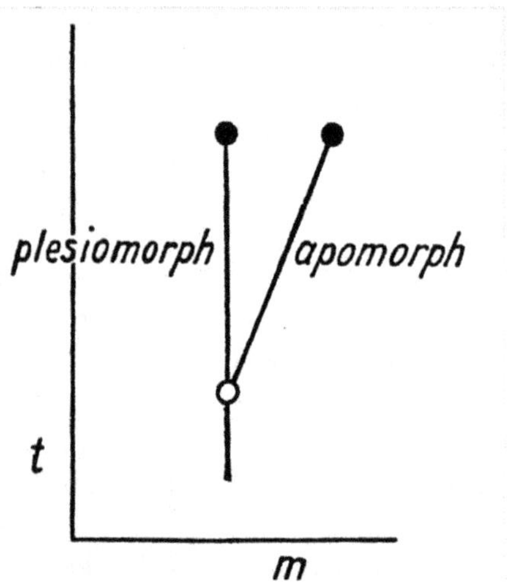

Fig. 65. Diagram illustrating ther meaning of the terms "apomorph" and "plesiomorph", from Hennig et al. (1953, no. 80).

He illustrated the meaning of 'apo-' and 'plesiomorph' in a graph that became widely known afterwards (Fig. 65). It resembles fig. 25 of the Grundzüge, but uses unambiguously the terms 'apo-' and 'plesiomorph'. At the end of the 1953 paper (no. 80), Hennig emphasised that "the age of a monphyletic group must not be equated with the origin of the morphological Bauplan that it realises today. Consequently, the age of a group as a tribe and the age of its typical Bauplan are not identical. Furthermore, one must not assume offhand that also the structuring of an old group must be of old age. 'Age of origin' and 'age of structuring' must, therefore, not be equated" ("das Alter einer monophyletischen Gruppe nicht mit der Entstehung des morphologischen Bauplanes, den sie heute verwirklicht, gleichgesetzt werden darf. Das Alter einer Gruppe als Sippe und das Alter ihres typischen Bauplanes sind also nicht identisch. Weiterhin darf nicht ohne weiteres angenommen werden, dass auch die Gliederung einer alten Gruppe ein hohes Alter haben muss. 'Entstehungsalter' und 'Gliederungsalter' einer Gruppe dürfen also nicht gleichgesetzt werden", p. 56). This differentiation denotes principally the same as the different ways to circumscribe monophyla described in the 1990s: 'apomorphy-based' (according to certain autapomorphies, i.e. the age of its Bauplan), 'stem-based' (with reference to a certain stem line, i.e. age of origin), or 'node-based' (referring to certain lineage splittings, i.e. its age of structuring) (De Queiroz & Gautier 1992).

Between 1955 and 1966, Willi Hennig demonstrated the application of his method repeatedly in taxonomic papers, mostly published in Beiträge zur Entomologie and—after 1961—in Stuttgarter Beiträge zur Naturkunde, always on Diptera. The only exceptions are the two small volumes on invertebrates of the *Taschenbuch der Zoologie* in 1957, 1959, 1963, and 1964 (nos. 96, 96a, 101, 101a). However, here Hennig did not give an explicit introduction in his method but rather silently applied his principles of division and used some of his terms quite inconspicuously, e.g. 'Grundplan' (which he had taken from Danser 1942).

Willi Hennig shortly took into consideration to replace the clumsy phrase 'monophyletic group' by the short and concise term 'monophylum' (no. 80: 9). Yet this name became accepted by a wider scientific community only after Ax suggested it again, in 1984 (p. 32).

In his contribution to the centenary of the DEI in 1956 (no. 95), Hennig published for the first time his later famous "Argumentierungsschema [argumentation scheme] der phylogenetischen Systematik" (Fig. 66). In this scheme, he exemplified how a phylogenetic-systematic hypothesis had to be substantiated: for each taxon, also for the terminal ones, at least one uniquely derived character (or character state) must be given, and for each apomorph character (or character state) the plesiomorph alternative must be indicated. It is remarkable how clear and graspable he wrote this paper, probably because it bases on a talk he gave on October 1, 1956. In the same paper, Hennig also published several illustrations, which he later used in his manuscript of *Phylogenetic Systematics*. This could mean that he worked already in 1956 on a revision of the *Grundzüge*.

From all these publication—and later ones—it is evident that Hennig's main aim was to group taxa by their relative degree of phylogenetic relationship, that is: according to the recency of the last common ancestor. What Brundin (according to Schlee 1971: 14) called "Hennig's principle" was the search for sister-groups, not for ancestors. Consequently, the question whether or not the stem species belongs to the same—monophyletic—taxon as its descendants, was not a major one for Hennig (but see chapter 5.4.).

Still until 1962, Hennig used "Restkörper" (remaining body) or "Restkörpergruppe" to denote the sum of all species remaining if one excludes one taxon or some taxa from a monophylum. Only in a report on changes in the phylogenetic system of the insects since 1953 (no. 108) in 1962, we find on p. 35 the term 'paraphyletic' as a denotation of such a group. In a respective footnote, Hennig stated that he had introduced this term "recently" ("kürzlich"). However, there is no earlier publication by him

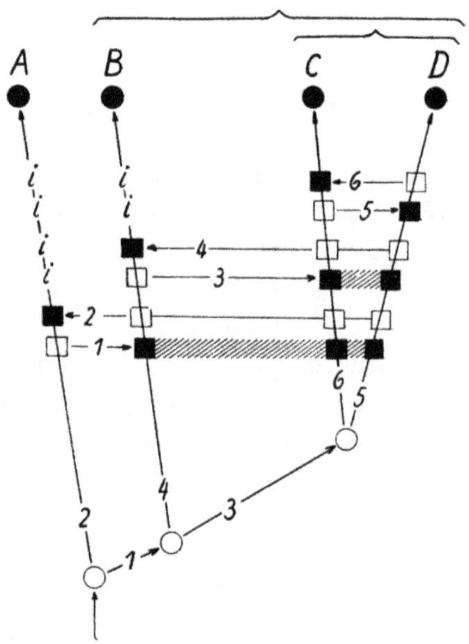

Fig. 66. The "argumentation scheme in Phylogenetic Systematics", from Hennig 1957 (fig. 9 in no. 95). Each of the putatively monophyletic taxa is characterised by the possession of at least one (syn-) apomorph character state, given in black. The sister-group of each monophyletic taxon has the corresponding plesiomorph state, given as open squares. The "i" indicates that the lineages A and B are "indifferent" as to the assessment of sister-group relationships between C and D or B and C+D, respectively. The numbers refer to the sections of the tree in which the respective character state evoloved, and the curly brackets on top represent the hypotheses on monophyletic taxa resulting from the distribution of character states.

where 'paraphyletic' could be found. Probably, this footnote refers to the manuscript of *Phylogenetic Systematics*, which he had submitted in 1961 already, but which appeared only in 1966.

5.2. Phylogenetic Systematics (*1966*) (Fig. 67)

Obviously, Willi Hennig already had completed a revised version of the *Grundzüge* at the DEI (i.e. before August, 1961). As Hull (1988: 134) described it, D. Dwight Davis of the Field Museum (Chicago) had contacted Hennig in 1960 as he was interested in an English translation of the *Grundzüge*. Hennig is said to have been "in the midst" of preparing a revised edition which he sent to Davis. Willi Hennig's eldest son

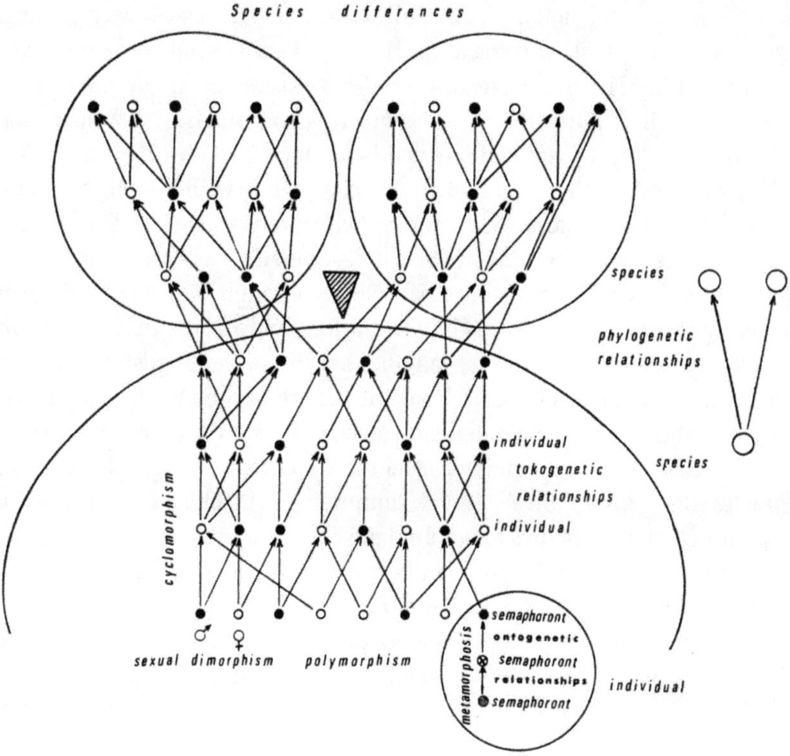

Fig. 67. The different forms of "hologenetic" relationships within and between organisms (fig. 6 from *Phylogenetic Systematics*—1966, no. 125). The different stages of an individual—"semaphoronts"—are connected by ontogenetic relationships; the different individuals within a species are connected by tokogenetic relationships, and between different species of a more comprehensive natural unit exist phylogenetic relationships.

Wolfgang published the original German version of this book posthumously in 1982. In his preface, he stated that his father had completed the manuscript in 1960 and that he had "neither information on the progress of the translation nor had he seen any version of it for checking" before the book was published in 1966, and that some minor parts of the manuscript had not been translated (no. 161: 5). Similarly, Hull (1988: 134) accuses the translators—D. Dwight Davis and Rainer Zangerl—that they "did not just translate the manuscript. They also heavily edited it, eliminating what they took to be repetitive passages, simplifying Hennig's Teutonic sentences, and clarifying his ideas", even that "Davis and Zangerl translating into English as equivalent to an anti-evolutionist translating Darwin into German" (Hull 1989: 5). As Hull writes, Dwight Davis was fluent in German but the task of translating Hennig's very special sentences overburdened him. He "appealed to one of his colleagues at the museum to help him, Rainer Zangerl, the Swiss-born paleontologist…". Zangerl had informed Hennig of the sudden death of Davis in February, 1965 (Doc. 128), which means that Hennig was not completely without contact to the translators. Nevertheless, Willi Hennig wrote to Gareth J. Nelson that "it may—by the way—be that in the English version… of my *Phylogenetic Systematics* not all of my views can clearly enough be seen by English-speaking readers. Perhaps it is important for you to know that I have submitted my German manuscript in 1961. I have seen the English text only in the printed book in 1966. I could not check it before the book was printed" ("Es mag übrigens sein, dass in der englischen Fassung meiner *Phylogenetic Systematics*… nicht alle meine Ansichten für Leser mit englischer Muttersprache klar genug zum Ausdruck kommen. Vielleicht ist es für Sie von Bedeutung zu wissen, daß ich mein deutsches Manuskript 1961 abgeliefert habe. Den englischen Text habe ich erst durch das gedruckte Buch 1966 kennengelernt. Ich habe ihn vor dem Druck nicht prüfen können", Doc. 129).

In my opinion, Davis and Zangerl did a completely fair job. Indeed, there are some differences between the books of 1966 and 1982, some minor, some major. The largest incongruence is most probably caused by an inadvertent slip: A rather long paragraph is missing on page 151/152 of the English book. There is an extensive discussion of certain characters of Brachiopoda in which the points are indented and labelled "(A)a", "(A) b", "(A)c" etc. The last six lines of "(A)c" of the German book are lacking, as well as the points "(A)d" through ("B"). This is especially odd because in the English text (2nd paragraph on p. 151) section "(B)" is explicitly mentioned. Other gaps are presumably not real "omissions". I suspect that Hennig modified his manuscript slightly after he had given it away in 1961. This would explain the existence of additional references, as in the first paragraph of p. 87 (English), additional paragraphs on p. 132 (English) about Haeckel and on extinction on p. 217 (English), or some more lines on reproduction in the 2nd paragraph of p. 234 (English), and especially a profound and long insertion on Fechtner about the "rule of progressive reduction of variability" that replaced a passage on chemical compounds on p. 218 (English). Assuming that Hennig worked on the manuscript after submission to Davies and Zangerl would also explain the different number of references: 304 in the English book, 312 in the German. When Wolfgang Hennig edited the German manuscript, these alterations could possibly not been recognised.

Several quantitatively minor changes modify to a certain degree the intensity of expressions, as on p. 194 (English), where the last paragraph begins with "It is often assumed...", whereas the German equivalent reads "Manchmal wird angenommen..." (sometimes it is assumed..., p. 189 German). I could only find extremely few instances where Hennig could possibly have protested, e.g. on p. 234 (English) we find "relationship" for the German "Zusammenhang" (p. 227 German) which should better read "context" in English.

In order to demonstrate how true Davis & Zangerl's translation reflects the German original, I reproduce a passage of p. 93 (English) or 96 (German):

(1) How does one determine which characters of different species must be regarded as transformation conditions (a, a', a", etc.) of one and the same character (Homology)? [1. Wie stellt man fest, welche Merkmale verschiedener Arten als Transformationszustände (a, a', a" usw.) ein und desselben Merkmals angesehen werden müssen? (Homologie)].
(2) How does one determine which is the beginning condition and which is the terminal condition of a transformation series (a, a', a", etc) (character phylogeny)? [2. Wie stellt man fest, welches der Anfangs-, welches der Endzustand einer Transformationsreihe (a, a', a" usw.) ist? (Merkmalsphylogenie)].
(3) Must the evolution of a character always progress in the same direction, or can a character revert to its original condition (a, a', a", a) (reversibility)? [3. Muss die Evolution stets in einer Richtung fortschreiten, oder kann ein Merkmal im Laufe seiner Transformation auch zum Ausgangszustand zurückkehren (a, a', a", a)? (Reversibilität)].
(4) Can certain transformation conditions of one character be attained by transformation of another character (a, a', a" and b, b', a") (convergence)? [4. Können bestimmte Transformationszustände eines Merkmals auch durch Transformation eines anderen Merkmals erreicht werden? (z.B. a, a', a" und b, b', a")? (Konvergenz)].
(5) Must a particular derived condition always be attained through a one-time series of transformation, or can the same condition be attained repeatedly from the same condition of origin (a, a', a" / a, a', a") (parallelism)? [5. Muss es stets eine einfache Transformationsreihe sein, durch die ein bestimmter Folgezustand erreicht wird, oder kann derselbe Zustand auch wiederholt, vom gleichen Ausgangszustand her erreicht werden (a, a', a" / a, a', a")? (Parallelismus)].

We can see that Davis and Zangerl tried hard to express Hennig's statements as correctly as possible, and in doing so they even accepted a certain oddness of their English. A thorough comparison of the two books revealed that the translation definitely conveyed Willi Hennig's ideas appropriately. This is the more remarkable since Dwight Davis and Rainer Zangerl were, as Hull (1989: 134) reported, "trained in the very philosophical tradition that Hennig attacked in his book—idealistic morphology". I could not find out why Hennig had no chance to check the English translation before the book was published. But it is clear that there would have been time enough—when the translation was completed, Hennig was established in Ludwigsburg and (in contrast to Hull's description) no longer "in the midst of fleeing from East to West Germany". In my opinion, Willi Hennig had no reason to complain about the English version of his book.

As a practical guide for character polarisation, Hennig offered several indices, all of them not really convincing: (1) geological character precedence: If one of two or more character states occurs in older fossils, and the other(s) in younger, then "obviously" the former is plesiomorph and the latter apomorph. Clearly, this "criterion" is only applicable if sufficient fossils are available, and this is rarely the case, especially in soft-bodied organisms. In addition, it is possible that the plesiomorph alternative is only known from younger sediments than the apomorph state, due to the incomplete fossil record. An incorrect polarisation would be the consequence (see Willmann 1989: 283). (2) Chorological progression: Apomorph character states occur with higher probabilities in the "apochor" daughter species, i.e. the species that "departed farther geographically or ecologically from the initial species". The problem is, however, how to identify this "apochor" species? It leads inevitably into circular reasoning if we aim at finding the species that "departed farther geographically or ecologically from the initial species" by searching for apomorph character states in their groundplan. (3) Ontogenetic character precedence: different character states occur during ontogeny in the same sequence as they evolved in phylogeny. Besides the fact that a systematist knows the ontogeny of the specimens he or she studies only in very rare cases, even Ernst Haeckel—to whom the "biogenetic law" is normally credited—warned that caenogenetic stages of development can disturb the picture. Distinguishing "caenogenetic" stages (those newly acquired in the evolution of a certain taxon) from "palingenetic" stages (those kept unchanged in a taxon from its ancestor) can only be done by applying the fourth "criterion". (4) Correlation of series of transformation. "Such correlations are of significance to phylogenetic systematics only if the direction in which one of two or more correlated transformation series is to be 'read' is known". This means, the interpretation of the yet unknown direction of transformation depends on decisions on other characters by means of one of the other three "criteria", which all are highly unreliable. Although Hennig gave numerous practical examples from the animal and the plant kingdom, the problem of how to polarise characters remained vague.

Hennig has certainly seen this. The solution he suggested is epistemologically obsolete (in my opinion) anyway: the so-called prinicple of reciprocal illumination, which I discuss below, in chapter 6. He had mentioned this principle in the *Grundzüge* already (on pp. 26, 201f., 272, 346) and stated that it was a proper way of reasoning.

Phylogenetic Systematics became a highly influencal publication in the field of phylogenetics. This can be estimated tentatively from the 4032 citations listed by Google Scholar (checked on October 9, 2012). A search under identical conditions revealed 20705 citations of Charles Darwin's *Origin of Species*, 7459 citations for Ernst Mayr's *Animal Species and Evolution* of 1963, 4431 for his *Systematics and the origin of Species* of 1942, and 281 for E.O. Wilson's *Sociobiology* of 1979/1980. Even the *Grundzüge* were cited 980 times, plus 109 citations of the Spanish translation. Of course, these figures must be interpreted with care, but they give an idea of the pace at which the scientific community accepted the new method.

5.3. *Mayr and the "Evolutionary Classification"* (Fig. 68)

In several of Hennig's replies to inquiries about an updated outline of his method he wrote that he did not like to write theoretical papers but thought the most efficient way to convince systematists of his approach was to demonstrate on concrete examples how to proceed (e.g. Doc. 154). On November 15, 1973, Hennig received a letter which made him change his mind, at least he agreed to write one more theoretical (rather: methodological) paper. Ernst Mayr had come to West Berlin on October 7, 1973. Erwin Stresemann, Mayr's fatherly friend and mentor, had died the year before, and Mayr visited Stresemann's widow Vesta Stresemann. He asked her to invite some of his old acquaintances for a round table discussion and a private chat. And so it came that Klaus Günther met Ernst Mayr again for the first time after 30 years. Mayr told Günther that he had authored a critique of Hennig's claim that a classification must reflect the results of a phylogenetic analysis one-to-one. He had written

Fig. 68. Ernst Mayr at the Weismann-Symposium in Freiburg im Breisgau, 29.05.1984, taken by Gudrun Mahlke.

this paper because some German colleagues, among them Otto Kraus and Wolfram Noodt, had met him on the 17th International Congress of Zoology in Monaco in 1972 and suggested that he could concisely outline his approach to classify, and that Willi Hennig could then reply in a similarly condensed manner (Doc. 155). Mayr agreed and sent his manuscript to Hennig on November 15, 1973. Hennig was reluctant, in the first place because he feared that an appropriate treatment of Mayr's critique would require a whole book. Finally, he wrote a shorter draft and sent it to Klaus Günther who commented on it in detail (Doc. 156). They sent the manuscript to Ernst Mayr and asked him to check it. He approved the text and had only one minor linguistic critique, as he had written in English, whereas Hennig's reply was in German: the correct translation of "confusing" is not "konfus" but "verwirrend" (Doc. 157).

In his paper, that appeared in 1974, Mayr did not criticise Hennig's method of analysing phylogenetic relationships. He rather emphasised that he accepted this method and regarded it as the first and necessary step towards an—as he called it—"evolutionary classification". The result of this first step is a hypothesis on the genealogical relationships among the taxa investigated, and this hypothesis reflects the underlying sequence of lineage splitting events, the so-called "cladogenesis" (Rensch 1947: 95). Accordingly, Mayr named Hennig's way of classifying "cladistic", a term coined by Cain & Harrison in 1960, based on the term "clade" of Julian Huxley (1958: 27). Mayr claimed that expressing phylogenetic (= genealogic) relationships as a result of a phylogenetic ("cladistic") analysis in a cladogram *and* in the classification would be redundant. Classification should be relatively independent of the phylogenetic analysis. After the analysis, the first step of classification, Linnean ranks should be assigned to taxa not exclusively according to their genealogy, but also "ecological role" or the "evolutionary success" of the taxa should be reflected. Consequently, sister groups would not necessarily be ranked equal.

An essential terminological and conceptual difference between Mayr and Hennig was the concept of "monophyly". Mayr referred to the origin of a taxon and insisted that also paraphyletic taxa have a single stem species. Therefore, he used 'monophyletic' to denote those taxa which stem from a single stem species and include it—in contrast to polyphyletic taxa which do not inlcude the last common ancestor of their members, a concept explained in detail by Mayr's colleague Peter D. Ashlock (1971). There, the term 'holophyletic' is used for Hennig's 'monophyletic', while the latter term should include 'holophyletic' and 'paraphyletic'. Mayr insisted that Hennig's usage of 'monophyletic' deviates from the 'traditional definition' applied ever since Haeckel (p. 104).

Hennig treated the question of inclusion or non-inclusion of the stem species explicitly only once. His verbal definitions always restrict the coverage of a monophyletic taxon to the species which have descended from a single stem species. In many of his illustrations, however, in the *Stammesgeschichte der Insekten* and in the *Stammesgeschichte der Chordaten*, the stem species is included within the circle or oval encompassing a monophylum. I found only one sentence in Hennig's publications that expresses this inclusion explicitly. In *Phylogenetic Systematics* (no. 125: 71) he wrote that "From the fact that in diagram I the boundaries of a 'stem species' coincide with the boundaries of the taxon that includes all its successor species, it follows that the 'stem species' itself belongs in this taxon" ("Aus der Tatsache, daß (in Darstellungsform I) die Grenzen einer 'Stammart' mit den Grenzen des Taxons zusammenfallen, das alle aus ihr hervorgegangenen Nachfolgearten umfaßt, geht hervor, daß auch die 'Stammart' selbst in dieses Taxon gehört", no. 161: 76). Surprisingly, he has never again considered this—certainly essential—modification of his own concept of 'monophyly'. Nevertheless, Hull (1970) expressed Hennig's view simply as "the stem species of every single higher taxon must be included in the taxon..." (p. 24).

Inclusion or exclusion of the stem species in a taxon matters because Hennig insisted in his reply to Ernst Mayr (no. 149) that paraphyletic and polyphyletic taxa do not differ with respect to genealogy since any two or more species must have a stem species in common (as long as the assumption of a single origin of the Metazoa, or the cell, or a single primary biogenesis, holds true). Rather their formation follows from two different types of errors: paraphyletic taxa are based on symplesiomorphies which are erroneously deemed to be synapomorphies, while polyphyletic groups are based on convergencies deemed to be synapomorphies. In contrast, Mayr argued that paraphyletic taxa—no matter how many subtaxa are excluded—and holophyletic taxa should be admitted in a classification because they are both "monophyletic" (sensu Ashlock), and that they are monophyletic because they both include their most recent common ancestor.

The example most fervently discussed in the two papers of Mayr and Hennig as well as in the scientific community are the reptiles. There is no doubt that "Reptilia" are not monophyletic, as the most recent common ancestor of all reptiles is also ancestor of the birds and possibly of mammals, too. Likewise, it is agreed that among the extant animals the crocodiles are most closely related to the birds. Mayr insisted anyways that in an "evolutionary" classification the reptiles had to be listed as a taxon—although not monophyletic—and ranked as a "class", like birds and mammals.

Mayr's fundamental thesis may be circumscribed as "Phylogenesis is cladogenesis, and evolution is more than that". In contrast, Hennig's position may be characterised by "Phylogenesis is history of phyla, and that is more than cladogenesis". He claimed that there is no methodologically faultless way to express in a system more than one *principium divisionis*. As long as we adhere to the postulation that all supraspecific taxa in a system must be monophyletic we cannot represent "evolutionary success" or anything else equally. To single out particular taxa and produce paraphyletic taxa on purpose, foils the crucial advantage of a strictly phylogenetic system, that is: all taxa are objectively founded and can be communicated unambiguously. There is an objective measure for phylogenetic relationship (the number of ancestors exclusively in common), but there is no objective measure for evolutionary success. Consequently, it would admit arbitrariness to allow this success being the criterion for the taxonomic rank and for subdividing the system.

Incidentally, neither Hennig nor other phylogeneticists in Central Europe restricted performing phylogenetic systematic to the reconstruction of the cladogenesis. Always they thought of the evolutionary history of the taxa when they polarised their characters, and in countless cases they also described it. The German-speaking phylogeneticists always considered at least part of the 'anagenesis', too. The latter term was defined by Bernhard Rensch in 1947 (p. 95) as "progressive evolution", a concept that—unnecessarily and unfortunately—the vast majority of biologists in Central Europe adopted. "Unnecessarily", because the theoretically decisive difference between cladogenesis and anagenesis has nothing to do with evolutionary progression. It is simply that the further denotes the sequence of lineage splitting events, while the latter means all evolutionary processes between species splittings. "Unfortunately", because the German wording for "progressive evolution" ("Höherentwicklung") implies the connotation of "higher" and "lower" taxa, and this is too often understood as "more valuable" and "less important", or "more complex" and "simple", while in reality the two terms just mean "more similar to humans" and "less human-like".

Kraus & Hoßfeld (1998) came to a similar judgment on the claim and the procedure of Phylogenetic Systematics: "However, the scientific approach of phylogenetic systematics is indeed much more comprehensive [than the reconstruction of the cladogenesis]. Criteria for the determination of the direction of the evolutionary change are acquired and applied, in terms of an integrative approach. Findings on functional morphology are considered, along with evaluations of 'evolutionary responses' to changing conditions of the environment" ("Tatsächlich ist jedoch der wissenschaftliche Ansatz der Phylogenetischen Systematik wesentlich weiter gefasst [als die Ermittlung der Kladogenese]. Im Sinne einer integrierenden Vorgehensweise werden Kriterien für die Bestimmung der Lesrichtung evolutiven Wandels erarbeitet und angewandt. Funktionsmorphologische Befunde, einhergehend mit der Einschätzung 'evolutiver Antworten' auf sich wandelnde Lebensbedingungen finden Berücksichtigung"). Walter Sudhaus (2007: 26) found a concise wording for this fact: "Phylogenetics is more than 'furcatology' ('Gabelkunde')".

In this context, a passage from a letter of Willi Hennig to Gareth Nelson (Doc. 197) might be enlightening. He wrote that the German noun "Verwandtschaft" is connected to the notion of "genealogical relationship" to a higher degree than the English word "relationship", and suspected that this might be one of the reasons why "we" (either he, or the German phylogeneticists, or both) can use 'Phylogenetic Systematics' without running the risk of misunderstandings, whereas (again) "we" do not like the denomination "cladistics".

Mayr's attitude was certainly not exclusively dominated by scientific motives but also by emotional ones. He was an ornithologist, and he was interested in the process of evolution. Thus, to rank birds and crocodiles equal, and to ignore "evolutionary success" in a classification was most probably for him not just difficult but mentally impossible. Mayr's younger follower and friend Walter J. Bock from Columbia University (New York, NY) made serious attempts to reconcile the lines of reasoning of Mayr and Hennig. In 1974, he wrote a manuscript of 59 pages on "methods of phylogenetic analysis" (Doc. 187) which he wanted to publish jointly with Willi Hennig and which he condensed to 13 pages in 1975 (Doc. 188). Bock concluded this offer to mediate by writing: "Distinct and different methods of phylogenetic analysis do not exist for phylogenetic classification and for evolutionary classification. It is our hope that the clarification of these methodological problems of phylogenetic analysis will reduce the scope of controversy between phylogenetic and evolutionary classification and permit focus upon critical issues" (Doc. 188: 13). Willi Hennig declined to be co-author (Doc. 189), because Bock had in his draft texts still argued that also paraphyletic taxa are "monophyletic" and should be admitted in a classification. For Hennig, it was simply unthinkable that someone had understood the method of phylogenetic analysis—Mayr and Bock definitely had understood it—but afterwards insisted in classifying differently from his own principles.

Gareth J. Nelson had offered to publish an English translation of Hennig's German reply to Ernst Mayr in *Systematic Zoology*. Thus, Willi Hennig sent his German manuscript to Nelson on August 15, 1974 (Doc. 161). Graham Griffiths supplied Nelson with a literal translation, which he "edited very carefully, with the purpose of achieving a clear and concise exposition in English" (Doc. 192). The English paper appeared in the June edition of *Systematic Zoology* in 1975 and did not evoke a noticeable discussion.

A remarkable misunderstanding demonstrates Hennig's poor command of English, and it is the more remarkable that neither Ernst Mayr nor Klaus Günther nor Dieter Schlee minded it although they knew Mayr's paper well and could compare Hennig's critique with the respective passages in the original text. Mayr had written (1974: 105) "No cladist seems to have noticed some of the consequences of the redefinition of the term monophyletic... Although starting from the entirely different premises, the cladist has methodologically returned to the 'divisional' method of classification that was dominant from Caesalpino to Linnaeus. His criterion of division is of course very different from that of the adherents of Aristotle's logical division, but the principle of classifying of both schools (cladists and logicans) is very much the same... It is the abandonment of the principle of upward classification, dominant since Darwin, and its replacement by Aristotle's downward classification which is the fatal flaw in the philosophy of cladistic classification". To this argument, Hennig refers by writing (no. 149a: 287) "Mayr must have felt that when he repeatedly refers to Aristotle and accuses the phylogenetic systematics that it had deviated from Aristotle's 'downward classification" ("Mayr muß das gefühlt haben, wenn er sich wiederholt auf Aristoteles beruft und der phylogenetischen Systematik vorwirft, daß sie von 'Aristotle's downward classification' abgegangen sei"). As Hennig had thanked explicitly Klaus Günther and Dieter Schlee for checking the manuscript, it is evident that also Günther and Schlee did not realise this error. Amusingly, the respective passage in the English version reads "Mayr must sense this because he repeatedly appeals to Aristotle, and reproaches phylogenetic systematics for adopting 'Aristotle's downward classification' (p. 105)". Hennig had read the translated text and thanked Gareth Nelson for his great effort (Doc. 193).

Since 1974, the feelings have calmed down remarkably, which is mostly due to the fact that the aim of phylogenetic (= cladistic) studies has been focused on the analysis of phylogenetic relationships, which means: production of cladograms, whereas ranking and naming taxa has lost importance significantly. This is also the impression of Quentin D. Wheeler, who wrote in an e-message "We . . . are seeing the neglect of attention to homology studies on the input side of phylogeny today as well as a neglect of formal classifications on the output" (Doc. 190).

5.4. *Later Works* (Fig. 69)

In the *Stammesgeschichte der Insekten* (posthumously published enlarged, commented, and updated as "Insect Phylogeny" in 1981 by Adrian C. Pont and Dieter Schlee) Willi Hennig discussed at length the problem of true and false stem-groups: true stem-groups comprise all fossils not belonging to the crown-group (this term was introduced by Jefferies in 1979), but to the pan-monophylum, whereas false ("unechte") stem-groups are made up also of taxa not belonging to the pan-monophylum (a term coined by Lauterbach in 1989). He treated in great detail questions of historical biogeography and palaeontology, e.g. plate tectonics and distribution of insect fossils of different geological age. His tree diagrams are attempts to visualise not just sister-group relationships but also geological age and species-richness. I find it remarkable that Hennig did not plot the putatively synapomorph characters on the lines of these diagrams. Instead, he explained verbally why he grouped certain taxa together or why he accepted them as monophyletic. It is certainly not just a bagatelle that he gave these characters in the form of simple statements: taxon so and so is apomorph in these and those characters and therefore monophyletic. He did not really discuss alternative interpretations, nor did he provide the foundation of his judgment. In many cases, he simply accepted the decisions of other authors—based on whatever considerations—without critical discussion, just relying on the competence of these authors.

In an attempt to overcome the limited number of Linnean hierarchical levels, Hennig introduced a system of rational numbering. The classification is expressed in series of numbers, e.g. "2.2.2.2..4.6..1." indicated the Mecopteroidea, on the next subordinate level "2.2.2.2..4.6..1.1." meant the Trichoptera, and "2.2.2.2..4.6..1.2." its sister-group, the Lepidoptera. He exemplified this method in a second case, in his manuscript on the phylogeny of the chordates (posthumously published by his eldest son Wolfgang in 1983).

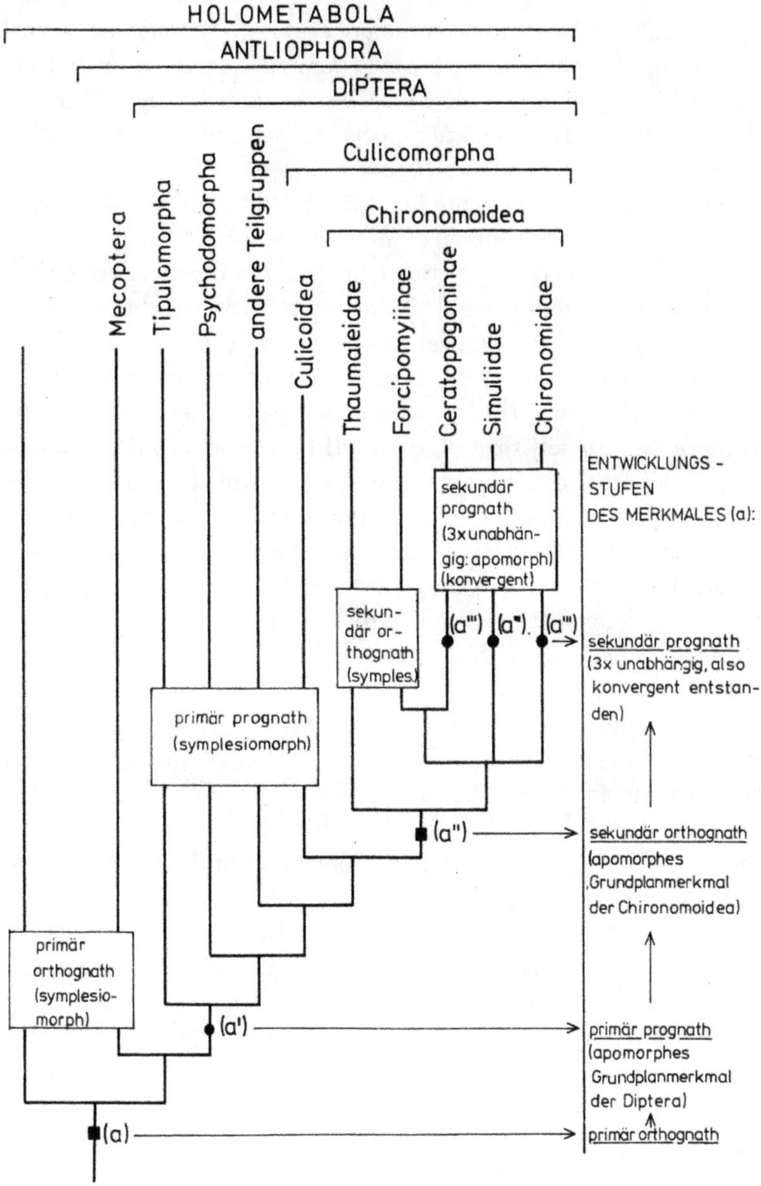

Fig. 69. Possible evolutionary pathway of orthognathy and prognathy in holometabolous insects, with special emphasis on the Diptera. Fig. 5 from Hennig 1984 (no. 167).

Obviously as a consequence of his discussions with Gerd von Wahlert (whom he acknowledged in the preface), he included ten pages on "phylogenetic research and evolutionary ecology". Although he begins by emphasising that phylogenetics would remain incomplete without evolutionary biology (p. 41), a closer look reveals that he only half-heartedly added here and there ecological aspects or discussed ecological prerequisites and consequences of an evolutionary step.

After completing the reply to Ernst Mayr, Willi Hennig could not publish any other papers on the method of Phylogenetic Systematics, but he had worked intensively on a further edition of the *Taschenbuch der Zoologie* and on some more general publications. They have been published posthumously by his son Wolfgang or by Dieter Schlee.

Towards later years of his life, Hennig's aptitude to write scientific texts improved considerably. The last publications penned by him are considerably more user-friendly than his early writings, especially the *Grundzüge*. He gave clear and concise overviews on his terms and definitions and provided plausible examples of their practical application. It is certainly a consequence of his sudden death, which prevented him from polishing his manuscripts, that some ambiguities remained. In the *Taschenbuch der Zoologie*, he defined "Grundplan" (groundplan) as 'all characters which have evolved in the hypothesised stem species of all the species of the considered taxon and only of them, which were passed on to the descendants either modified or unchanged' (no. 96d: 24). This means that those characters of the stem species that had an earlier origin are not included in the Grundplan. In contrast, he stated in the *Stammesgeschichte der Chordaten* (no. 160: 9) that "plesiomorph characters (groundplan characters, not *derived* groundplan characters) can not define monophyletic groups" (emphasis in the original).

Anyway, he had defined the terms explicitly in a glossary on the first 13 pages of the *Taschenbuch der Zoologie* (no. 96d), which are of great value for students (but useful even for professionals who want to make sure that they have adopted Hennig's thoughts exactly). The same is true for the introduction of the *Stammesgeschichte der Chordaten*, and especially for the little book on *Aufgaben und Probleme stammesgeschichtlicher Forschung*. As already mentioned, Hennig made clear in this publication (no. 167: 41) that 'plesio-' and 'apomorph' refer to hypotheses on characters, not on taxa. This clarification matters a lot, because many German phylogeneticists of those days did (and some perhaps still do) polarise characters by reference to the status of their bearers: A certain state must be plesiomorph because it occurs in taxon A, and this taxon is the most primitive one. This method corresponds to the "criterion of the correlation of transformation series", but it leads to a circular argument—a taxon is the most primitive, because most of its characters are plesiomorph, but this state had to be assigned only after the process of which it is now a premise.

In addition, he explained in detail (as in the *Taschenbuch der Zoologie*, no. 96d) how he saw the relation between 'synapomorphy' and 'homology' (pp. 37ff.). He gave the definition: "Homologous characters are characters, which evolved—modified or unchanged—from one character in the common stem species of their bearers" ("Homologe Merkmale sind Merkmale, die verändert oder unverändert aus einem Merkmal der ihren Trägern gemeinsamen Stammart hervorgegangen sind"). Like in 1953, he emphasised that the plesiomorph and the apomorph alternatives (states) of a character are homologous. Mere assessment of 'homology' is, in his view, however not sufficient in phylogenetic research, for at least two reasons: (1) The notion of 'homology' does not imply the direction of evolutionary change—the statement "the bird wing is homologous to the monkey forelimb" is as correct as the opposite. (2) 'Homology' applies without restraint only to structures, i.e. "positive characters". 'Character', however, can also mean "lack" or "loss" of a structure. In traditional phylogenetics, such a "negative" character cannot be identified by any of the conventional homology "criteria".

An interesting and most striking fact is that he spent two paragraphs (p. 46f.) on the argument that it is—in his opinion—impossible to provide a general method of character polarisation. As several authors have recognized (e.g. Hull 1988: 130; Kitching et al. 1998; Richter & Meier 1994; Schmitt 2001: 333), Hennig put remarkably little effort into elaborating a generally applicable and methodologically unobjectionable method for polarising character states. Even his publications after 1966 remained vague and hardly useful for practical purposes in this respect. Also in personal letters (as kept in the archive of the SMNS), he denied even the possibility of providing a general tool for assessing character polarity. In the *Aufgaben und Probleme* (no. 167), he even wondered that it is "strange to say that sometimes the phylogenetic systematics is blamed for failing to provide or develop methods for the discrimination of apomorph and plesiomorph characters" ("merkwürdigerweise wird der phylogenetischen Systematik manchmal zum Vorwurf gemacht, daß sie es versäumt habe, Methoden anzugeben oder zu entwickeln, mit deren Hilfe plesiomorphe von apomorphen Merkmalen ... unterschieden werden können", p. 46f.). It is, in my opinion, highly irritating that Willi Hennig could wonder why someone expected him to do the most plausible and useful thing in the given situation. On p. 47, he explicitly stated that it is equally impossible to offer a simple method or infallible criteria for the assessment of 'homology' as for the decision which of two homologous character states are apomorph or plesiomorph.

Instead of offering a general tool or method, he referred in this booklet to lists of primitive and derived characters for insects to be found in literature (p. 47). Only in a joint paper with Dieter Schlee, posthumously published in 1978 (no. 162), did he write (or agree to) some explicit statements on this topic. Astoundingly, Hennig & Schlee thought that the distinction of 'apomorph' and 'plesiomorph' can mostly be achieved quickly ("Die meist schnell gewonnene Unterscheidung zwischen plesiomorpher und 'abgeleiteter' Merkmalsalternative"), and that the following separation of uniquely derived (= 'synapomorph') characters from convergent or parallel needs special considerations. Among them comes "distribution of the characters among the taxa" closest to what is now known and used as "outgroup comparison method" (e.g., Kluge & Farris 1969; Wiley 1981: 139; Watrous & Wheeler 1981).

Worth mentioning is the emphasis Hennig laid on two books by Rolf Löther (1972a, b), already in 1974 (no. 149), but especially in 1984 (no. 167). This is remarkable because Löther is avowedly a Marxist philosopher, and Hennig was definitely an Anti-Communist. The background is that Hennig had slowly acknowledged that 'dialectics' is not an invention of the Marxists but dates back to the cradle of Western culture, Ancient Greece, and to one of (South-)Germany's most renowned philosophers, Georg Wilhelm Friedrich Hegel (Doc. 132). In addition, he realised that Phylogenetic Systematics became widely, and in a way "officially" accepted in the GDR earlier than in West-Germany (see Peters 1971a, b; Königsmann 1975; Peters & Klausnitzer 1978). Löther's books confirmed that Phylogenetic Systematics plays an indispensable role also in the frame of a Materialistic view of science (Schmitt 2010b). I find it most remarkable that in the GDR, as shown by the cited papers by Günther Peters, the method of phylogenetic systematics was a topic in grade 11 at school ("Oberschule"), already in 1971.

The later in his life the more had Willi Hennig the impression that he received less public appreciation than he deserved (e.g. Doc. 198), especially after Ernst Schüz had retired as the director of the SMNS in 1969. In 1966, he had written to Klaus Günther (Doc. 198): "If I read the many reviews of my papers from many years, and if I note that my nomination as a member of the Leopoldina 'on the basis of proposals from both parts of Germany' should express how 'great my achievements in the field of systematics and morphology especially of the Diptera' are, and that by this nomination should be acknowledged 'in particular' that I have aimed 'to comment beyond my special field of work on the most general problems of biology', then I feel in a way ashamed that I did not reach more than many others who were content to write Kosmos-booklets and the like or to discover some 50 new species of tipulids" ("Wenn ich mir die vielen Besprechungen meiner Arbeiten aus vielen Jahren durchlese und auch zur Kenntnis nehme, dass mir mit meiner Ernennung zum Mitglied der Leopoldina 'auf Grund von Anträgen aus beiden Teilen Deutschlands' zum Ausdruck gebracht werden sollte, wie 'groß meine Leistung auf dem Gebiete der Systematik und Morphologie, insbesondere der Dipteren', ist, und daß damit 'in besonderer Weise anerkannt' werden sollte, daß ich 'über meine Spezialarbeiten hinaus versucht habe, zu den größten allgemeinen Problemen der Biologie Stellung zu nehmen', so schäme ich mich eigentlich etwas darüber, daß ich damit nicht mehr erreicht habe als viele andere, die sich damit begnügt haben, Kosmosbändchen und ähnliches zu schreiben oder einige 50 neue Tipulidenarten zu entdecken"). We see that Hennig constructed long-winded sentences even in private letters. After all, to realise that his phylogenetic method had found its way into the acccepted academic curriculum must have brought a degree of satisfaction. Understandably, Hennig cited Löther's books extensively on pp. 8–13 in the *Aufgaben und Probleme*, and he used him as principal witness when he critisised the Numerical Taxonomy (on p. 24.).

5.5. *Hennig and Cladistics* (Fig. 70)

What today is widely accepted as "cladistics" is definitely not identical with Willi Hennig's original concept of "phylogenetic sytematics". The term 'cladistics' was coined by Ernst Mayr in 1965 and refers to 'cladistic' (Cain & Harrison 1960) but stems from Julian Huxley's term "clade" (1958: 27). Hennig did—at least originally—not like the idea to be a "cladist". He wrote to Gareth J. Nelson (Doc. 197) that "cladistics" has a slightly pejorative connotation ("ein leichter Beigeschmack von Abwertung"). In the *Taschenbuch der Zoologie* (no. 96c, 96d: 23) he wrote that the opponents of (his) Phylogenetic Systematics apply the term 'cladistic' to it in a derogatory way. Therefore, we can firmly assume that he would not have appreciated the historical development of the past 30 years or

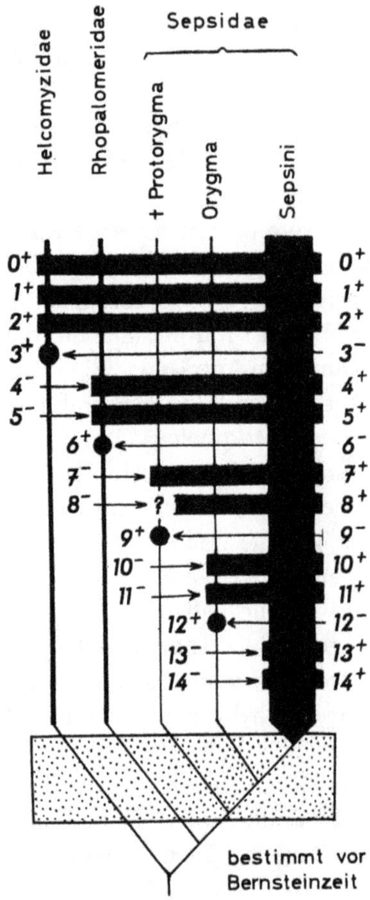

Fig. 70. Concrete example of an argumentation scheme, from Hennig (1965, no. 118, Fig. 78).

so that made 'cladistics' the nearly universally accepted label for Phylogenetic Systematics (as opposed to Evolutionary Taxonomy).

Hennig's ideas are probably most accurately and congenially transferred into English by Wiley (1981, 2nd edition: Wiley & Lieberman 2011). According to Dupuis (1984), they—merely—form the foundation of present-day cladistics. Since Hennig's untimely death in 1976, the phylogenetic method developed by Hennig has undergone at least two major modifications:

(1) The elaboration of the out-group comparison method. The term 'out-group' was first used by Throckmorton (1968), but the concept of

out-group comparison has been developed by Kluge & Farris (1969), and later formalised by Wiley (1981: 139) and Watrous & Wheeler (1981): if a character in question occurs in the ingroup—the group of taxa analysed—in two states, say a1 and a2, and in a taxon which does definitely not belong to the ingroup—which is then called the outgroup—only in one of these, e.g. as a1, then it is more parsimonious to regard a1 as plesiomorph and a2 as apomorph. It is important to note that 'outgroup' denotes nothing but the role of a taxon in the analysis. Demanding that the 'outgroup' has to be the 'sister group' would lead to an infinite regress.

Hennig—and many others—had used this method implicitly (their "criterion of the distribution of characters among taxa") but did certainly not consider this "criterion" to be the most useful one. The outgroup comparison method reached high impact through its implementation in computer programmes based on parsimony algorithms (e.g. Farris 1988; see Forey et al. 1992). The most important consequence of applying these computer programmes is, in my opinion, that polarisation is not longer done for each character separately (which means in the ideal case: independently) but by labelling one or some of the rows of the data matrix 'out-group(s)'. Consequently, there is no longer need for a priori reasoning over the possibilities of evolutionary transformation, i.e. before the tree calculating programme is executed.

(2) The claim to separate strictly the analysis of the pattern formed by the distribution of the characters analysed through the taxa under consideration from the explanation of this pattern by reference to evolutionary processes by the so-called 'transformation of cladistics' (Nelson 1979; Platnick 1979; Patterson 1980). Although there are certainly good arguments for such a strict separation, it was and is not accepted throughout, especially not in Willi Hennig's country (e.g. Ax 1984: 40). As Hull (1989: 12) pointed out, pattern cladists have departed from Hennig in several respects, up to the total reversion of one of Hennig's original views (concerning the logical and historical precedence of a non-phylogenetic system over the phylogenetic one, a position held by Platnick, 1985, but deemed "absolutely wrong" by Hennig, no. 76: 15, no. 125: 11).

To be certain: I do not regard "pattern cladism" *in toto* as a major success. I just think that the emphasis on the difference between pattern and process was important and useful. Hennig himself, and

> countless of his followers, polarised characters by stressing *ad hoc* arguments pertaining to ideas on the evolution of the treated taxa. A common pattern of reasoning was, e.g., that the last common ancestor of Taxon A had an aquatic mode of life, consequently all adaptations to a terrestrial life must be apomorph. Hennig followed such a way of argumentation in many discussions of characters in his *Stammesgeschichte der Chordaten*, e.g. when treating the origin of the Tetrapoda or of the Amniota.

To a great extend, the overwhelming success of Hennig's method is due to the fact that it can be formalized and, therefore, executed by a computer. This is also true for the parsimony based "Wagner groundplan divergence method" (Wagner 1961). This method had "received ... no attention from zoologists until Kluge and Farris (1969) used Wagner's basic method to produce a phylogenetic computer algorithm ..." (Wiley 1981: 176f.). The "grand alliance between the Wagner tradition and the Hennig tradition" (E.O. Wiley, personal communication 2001) formed the basis for the revolution of systematics during the last three decades of the 20th century. Wagner had given a concise instruction how to work out a phylogenetic problem in three phases: "(a) systematic or comparative analysis of the plants in question to find and understand their contrasting characters; (b) determination of ground plans to find the character states common to all or most of the plants in order to deduce the most probable ancestral or primitive states; and (3) phylogenetic synthesis to assemble the taxa according to their respective deviations from the basic ground plan and from each other". Furthermore, he described five "detailed steps" of the practical analysis: "(1) to compare and study all the variable characters among the taxa; (2) to determine the generalized or primitive conditions on the principle that characters found in most or all of a number of related taxa are inherited essentially unchanged from the common ancestor ...; (3) to assign for each character the value 0 for the generalized or primitive condition, and 1 for the specialized or secondary condition ...; (4) to list in a tabular form the taxa and for each give the divergence values from the ground plan, both for individual characters and in total; and (5) to determine the mutual character groupings between taxa and then arrange them in sequence according to these groupings on a concentric chart or graph, the radii and branchings to be determined by the mutual character complexes, and the distances by the divergence indices. So that the facts may be made readily visual, the secondary or advanced states of each character should be expressed by letters ...".

It is evident that this procedure is much more practical than Hennig's guidelines, and it is clear that Wagner's "detailed steps" provide an excellent opportunity to implement them in a computer programme. Indeed, James S. Farris published a "method for computing Wagner trees" in 1970. Why Hennig seemingly did not take notice of Wagner's or Farris' papers, is entirely mysterious. In the *Grundzüge* as well as in *Phylogenetic Systematics* he definitely considered botanical papers or papers by botanists (e.g. Zimmermann 1937; Schwanitz 1953), and in *Phylogenetic Systematics* he cited twenty papers from *Systematic Zoology*, where Farris' paper had appeared. This means that he either had overlooked it or not realised its relevance.

The affinity of Wagner's method and the "Numerical Taxonomy" outlined by Sokal & Sneath is striking. Indeed, James S. Farris was a follower of this school of systematics—named "phenetics" by Ernst Mayr in 1965—in 1970 and "converted" only later—about 1974—to "cladistics" (according to Hull 1988: 155; 1989: 6). Hennig had met Robert Sokal at the occasion of the 12th International Congress of Entomology in London in 1964, and he found him "really likeable" ("recht sympathisch") (Doc. 200). He even ordered Sokal & Sneath's book *Principles of Numerical Taxonomy* for the museum library. This means, at least in the beginning he did not strongly oppose "Numerical Taxonomy". Therefore, it is well possible that he would have appreciated the numerical methods developed in a 'cladistic' framework, especially by Farris (e.g. 1988), starting actually from Wagner's ground plan divergence method, but they could as well have started from Hennig's principle (Farris 2012).

It is, of course, futile to discuss the probability that Willi Hennig would have liked or disliked papers he never could read. It may, however, not be completely useless to speculate on that question in order to decide to what degree "cladistic and phylogenetic analysis today" (Platnick 1989) represents Hennig's original ideas. Considering Hennig's aptness to invoke mathematics and statistics in his early papers, and also in the *Grundzüge* (however less so in *Phylogenetic Systematics*), he may have joined the 'cladistics', especially the 'numerical cladistics' party. I dare to state, though, that he would never have accepted the method of polarising character states by rooting an unrooted tree. He was so thorough in describing and discussing each putatively synapomorph character, and he emphasised so energetically the importance of finding arguments for the assignment of the direction of transformation that we can firmly assume that he would have insisted in polarising each character separately. In addition, the recognised frequency of "specialisation crossings" ("heterobathmy of characters") would certainly have made him hesitate to define one single species, or even several concrete taxa, as possessing all characters of the matrix—or at least those characters with different states in the ingroup and in the outgroup—in the plesiomorph state. However, the "outgroup comparison method" is more complex than just that, and lastly it bases upon Occam's razor. Parsimony is certainly a criterion that Hennig would have accepted, as he had applied it so often himself, albeit not labelled as such.

Hennig's way to polarise characters independently and individually, and thereby also using adaptational arguments, is possibly the decisive difference between 'cladistics' (at least as long as this means 'transformed cladism') and traditional, i.e. Hennigian, phylogenetics (see Wägele 2004).

Some major flaws and weaknesses of Willi Hennig's phylogenetic systematics cannot be neglected. First, as already mentioned, he declined to provide a general empirical tool for character polarisation. The "criteria" he gave in 1966 (paleontological order, ontogenetic sequence, geographic distribution, and correlation of transformation series) are either theoretically obsolete (the last) or they can only be applied in special cases. This disadvantage was only overcome by the method of out-group comparison (e.g., Kluge & Farris 1969; Wiley 1981; Watrous & Wheeler 1981). Second, he was rather vague concerning the relation between his concept of synapomorphy and the traditional homology concept. This led to a considerable degree of confusion and to views far apart from Hennig's own ideas, which were so close to traditional continental European usage that he most probably saw no need for clarifying his position explicitly in the *Grundzüge*. Third, Willi Hennig never gave up the idea that a—well substantiated—phylogenetic hypothesis should be converted one to one into a taxonomic classification. Here, necessarily a conflict arose between the limited number of taxonomic ranks and the discovered number of splitting events, each of them would require a separate rank for the resulting taxa. Hennig tried to cope with this problem in his *Stammesgeschichte der Insekten* (1969, no. 138) but his solution (superlong sequences of numbers) was not convincing. Another irritating aspect is the high frequency of nomenclatorial changes, according to the publication of new phylogenetic hypotheses. These unsolved problems stimulated unconventional suggestions, e.g. consequent renunciation of Linnean ranks and invariant use of current names (De Queiroz & Gauthier 1992), but also caused a good portion of systematists to refrain from formally classifying but restricting themselves to produce cladograms.

Seemingly, Hennig hesitated to accept the term 'cladogram'. In his "Anti-Mayr paper" (no. 149), he put it between quotation marks and used it interchangeably with "phylogenetic tree" and "classification". Also in later publications he continued to call the schematic drawings he presented "diagram of relations" ("Verwandtschaftsdiagramm") or "phylogenetic tree" ("Stammbaum"). Thus, it remains uncertain if he ever had accepted the statement of, e.g., Platnick (1977) that cladograms and phylogenetic trees are not the same and must not be confused.

Hennig presented cladograms by far not in all of his papers after 1950. To a greater extend he did so only in his *Stammesgeschichte der Insekten* (no. 138). Even when he discussed hypotheses of the phylogenetic relationships of the taxa he had studied, he expressed his arguments in most cases only verbally. Three papers on the fly fossils from the Lower Cretaceous are remarkable exceptions (nos. 118, 142, and 145). Here, he followed his own principles most strictly and indicated the characters he regarded synapomorph in a clear way graphically on cladograms. However, the schematic diagrams he normally presented indicate just the sequence of lineage splitting events, in some cases the geological era in which these events might have taken place, and in some rarer cases also biogeographic relationships. In earlier publications (e.g. no. 15), but also in 1953 (no. 80) and in 1969 (no. 138), he intented to indicate the number of species of the taxa under study by the diameter if the lines representing them. However, he never attempted to express aspects as "level of complexity", "degree of (overall-)apomorphy", "adaptational success", and the like. Therefore, when looking at the graphs exclusively, one could get the impression that Hennig was only interested in the most parsimonious way to plot character state distributions over taxa, which would be evidence of a profound affinity to the "transformed cladistics" that evolved only after his death.

Fig. 71. Willi Hennig on vacations in Northern Italy, 1965. Courtesy of Gerd Hennig, Tübingen.

CHAPTER SIX

THE PHILOSOPHER

Willi Hennig made considerable efforts to found his "theory" of phylogenetic systematics on safe epistemological grounds. During his student times, he received professional tuition in philosophy only by Arnold Gehlen in but one seminar. However, it is evident from his letters to his wife (especially Doc. 049) and his correspondence with Klaus Günther that he was highly interested in philosophy. Consequently, he referred in the *Grundzüge* to Max Hartmann, Nikolai Hartmann, Theodor Ziehen, and indeed also to Gehlen. He cited, e.g., Ziehen for a term as basic as "order", obviously to prevent critique from pedantic opponents. One concept that Hennig invoked in a prominent manner is the "method of reciprocal illumination", on which numerous recipients of his "theory" commented, e.g. Sokal & Sneath (1963) and Hull (1967). Hennig stated that this method is "known in humanities" (no. 76: 26). He took this principle from an ethnologist, Wilhelm Emil Mühlmann, who reported that by application of this method, an ethnologist can understand a foreign culture like that of a Melanesian people, by studying a sector of that culture, e.g. arts, in the light of his knowledge about another sector, e.g. religion, gains a deeper insight ("Verständnis") of arts, which in turn leads to a deeper insight in religion, and so on. This procedure has a rather trivial side—of course, any new data will increase our knowledge. Of course, it is appropriate to re-check a former source of information when new information is available. Insofar is the method of "checking, correcting, and rechecking of the Anglo-Saxon authors" (no. 125: 21) nothing new in science, as Hull (1967: 186) stated, and Bock underlined (1968: 647): "It is a common method used by systematists and evolutionists but rarely acknowledged clearly". There is, however, a serious flaw related to it. If this method is logically circular, as Hennig conceded in the *Grundzüge* (76: 26), then it cannot yield logically correct arguments. It can be used in order to invent hypotheses, but never in order to test hypotheses. Moreover, since the method was established in the humanities explicitly as a tool to understand ("verstehen"), the question arises if there is anything in the order of nature that can be "understood" in the same way as the meaning of arts in a culture can be understood. There is some irony in the fact that Willi Hennig transferred a method from humanities to science and stressed it several times at important positions (no. 76: 201f., 272, 346), and on the other hand warns again and again (e.g. 125: 76, 213) against the logical error of a *metábasis eis allo génos* (a transgression to another field). It could well be that he committed just that.

The debate on the ontological status of species as individuals or as classes, e.g. in Hull (1976), Wiley (1980), Bunge (1981), Bernier (1984), Ghiselin (1987, 1989), de Sousa (1989), Stamos (1998), Rieppel (2007b, 2010)—to name just a selection—is not an invention of the past fourty years. Hennig devoted several pages to the question whether or not species—and supraspecific taxa—are individuals, or at least individual-like kinds. He refers to Theodor Ziehen, Nikolai Hartmann, and Ludwig von Bertalanffy for a definition of "individual" (no. 76: 114ff.). On p. 116, he concludes preliminarily that the higher taxonomic units as well as the species and the individuals *sensu stricto* possess, in a way, individuality and reality. However, he recognises principal—not merely gradual—differences between the "conventional" individuals and the supraindividual units. On p. 120, he writes that also the "higher groups" in a phylogenetic system are 'individuals', however in a restricted sense. This fact is, as he states, extraordinarily important for biological systematics. On p. 297, he explains that—as the higher taxonomic units possess individuality, even if in a restricted sense—any change in a "conventional" individuum, and even more so in a species, is a contribution to the history of the superordinated taxon. Towards the end of the last chapter, we find somewhat mystical considerations on the question if those "individuals of higher order" can develop a "Gruppengestalt", and if so, if this supraspecific "Gestalt" can further evolve. Although Hennig put these thoughts very cautiously, it is obvious that he sympathised with this idea.

In spite of his interest in philosophy, Willi Hennig obviously never referred to Karl Popper's *Logik der Forschung* (The Logic of Scientific Discovery) although he owned a copy of it (Doc. 205). Consequently, he did not comment on him. Therefore, it is not known if he would still have sought for "laws" and "rules" of evolutionary change if he had taken notice of Popper's views. Great parts of the *Grundzüge* and still of *Phylogenetic Systematics* are filled with considerations of laws and regularities in evolution which could yield tools for character polarisation. It is in this context that Hennig cited Dollo, Rensch, Rosa, Sewertzoff, Zimmermann and many other authors who had earlier treated this issue. At any rate, nothing in Hennig's writing indicates that he had, as would have followed from Popper's philosophy, accepted that there are no laws guiding history, nor evolution. As Stephen J. Gould put it drastically on October 10, 1996, during a symposium on "Organisms, Genes, and Evolution—Evolutionary Theory at the Crossroads", held in Frankfurt am Main (Germany): "There are no laws in history—it is just one damn thing after the other". Since Willi Hennig was obviously focused on a lawful evolution (see Rieppel 2007a), this view possibly prevented him from searching for an operational tool for polarising characters.

On the other hand, Hennig saw sharply that there is no epistemological justification for the distinction between "nomothetic" and "idiographic" sciences. These terms date back to Windelband (1904: 12), who had introduced them in order to denote the difference between sciences and humanities. Hennig stated on p. 2 of the *Grundzüge* that Windelband's distinction is generally useless for the division of sciences/humanities. It denotes rather two different points of views under which sciences and humanities likewise must be carried out. This idea resembles strikingly the concept outlined by Ernest Nagel eleven years later (Nagel 1961). Nagel points out that historical explanations have to rely on deductive conclusions from universal statements, and scientific explanations contain reference to singular statements. For Hennig, it bore special relevance to emphasise the unity of nomothetic and idiographic sciences. Systematics, as he saw it—i.e. a phylogenetic systematics—was regarded an idiographic enterprise (in German 'Wissenschaft' comprises 'science' *and* 'humanities'). If there would be a strict separation between nomothetic and idiographic 'Wissenschaften', then systematics would be necessarily non-scientific. It was his main concern to explain that the placement of systematics within science is fully and theoretically correctly justified.

Hennig's desire to make systematics a science proper is also evident from his emphasis on the power of systematics to provide explanations. He agrees with Naef (1919: 74) that phylogenetics seeks for explanations "like all sciences" ("Naturwissenschaften"). According to Windelband, the aim of science ("Naturwissenschaft") is to explain, whereas the aim of humanities ("Geisteswissenschaft") is to understand. Thus, the claim that phylogenetics can provide explanations (no. 76: 201), is crucial for Hennig's fundamental view of his own practice. At this stage, he struggled not only for his self-esteem but, moreover, for the reputation of a whole discipline.

Hennig's practice to represent his phylogenetic hypotheses, whether they refer to concrete taxa or to general models, in strictly dichotomous graphs, "cladograms", raised again and again discussions on the question if species split obligatorily into two descendants, if Hennig saw it like that, and if he invoked a lawful eplanation for this statement.

Hennig had indeed considered natural causes for the observation (!) that in *many* cases (my emphasis) a phylogenetic analysis of taxa of the phylogenetic system reveals exactly two subtaxa (no. 76: 332). It is also clear that he regarded a dichotomous structure as "by no means unlikely" ("keineswegs unwahrscheinlich", p. 333). However, on several occasions he clearly stated that he did not assume a logical or natural necessity for dichotomies. Thus, he explained that he started from assuming two daughter species "for simplicity's sake" ("der Einfachheit halber", p. 352). In *Phylogenetic Systematics*, we read on p. 210 "If phylogenetic systematics starts out from a dichotomous differentiation of the phylogenetic tree, this is primarily no more than a methodological principle". This is entirely plausible, because some basic thoughts show irrefutably that a dichotomy proven by the detection of at least one autapomorphy for the terminal taxa and at least one synapomorphy for both together can hardly be anything else. A trichotomy or any other polytomy, however, can never be proven definitely, thus it can always "in reality" be a sequence of dichotomies which cannot be resolved simply due to insufficient data.

It may well be, as Rieppel (2011b) described it, that "Hennig's realism required an isomorphism between nature and the natural system he was seeking", and that he "consequently could not avoid the ontological grounding of the 'principle of dichotomy' in speciation: speciation is dichotomous and nothing else". Considering Willi Hennig's own writings on this "principle", I think the question whether or not he "grounded" it ontologically is more a matter of psychology than of philosophy. If "Hennig's Principle"—the quest for sister-groups—means always to regard your analsysis incomplete as long as you cannot present a "completely resolved", i.e. strictly dichotomous, tree, you may easily be attempted to assume that all real trees indeed must be bifurcated only. Nevertheless, since Hennig nowhere stated that species split obligatorily into only two daughter species, nor that trees must necessarily be structured dichotomously, we should be fair and refrain from insinuating a certain type of philosophy. My impression is that Willi Hennig was interested in philosophy, but that he was not really a trained philosopher. That he referred to Wilhelm Dilthey in the *Grundzüge*, but had Hans-Georg Gadamer's *Wahrheit und Methode* on his bookshelf does certainly not allow assigning him either to the "traditional hermeneutics" of the first or to the "universal hermeneutics" of the latter. He certainly did not purposefully join a certain school of philosophy, but rather eclectically took ideas from here and there.

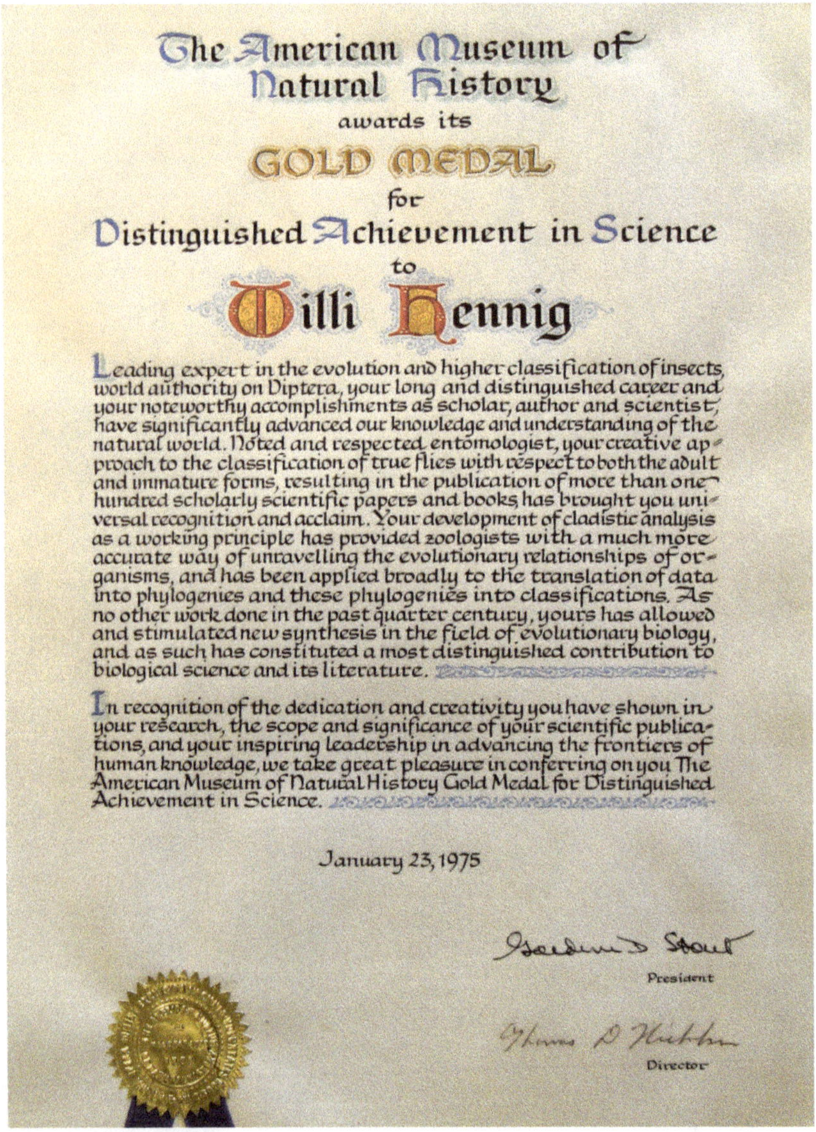

The American Museum of Natural History
awards its
GOLD MEDAL
for
Distinguished Achievement in Science
to
Willi Hennig

Leading expert in the evolution and higher classification of insects, world authority on Diptera, your long and distinguished career and your noteworthy accomplishments as scholar, author and scientist, have significantly advanced our knowledge and understanding of the natural world. Noted and respected entomologist, your creative approach to the classification of true flies with respect to both the adult and immature forms, resulting in the publication of more than one hundred scholarly scientific papers and books, has brought you universal recognition and acclaim. Your development of cladistic analysis as a working principle has provided zoologists with a much more accurate way of unravelling the evolutionary relationships of organisms, and has been applied broadly to the translation of data into phylogenies and these phylogenies into classifications. As no other work done in the past quarter century, yours has allowed and stimulated new synthesis in the field of evolutionary biology, and as such has constituted a most distinguished contribution to biological science and its literature.

In recognition of the dedication and creativity you have shown in your research, the scope and significance of your scientific publications, and your inspiring leadership in advancing the frontiers of human knowledge, we take great pleasure in conferring on you The American Museum of Natural History Gold Medal for Distinguished Achievement in Science.

January 23, 1975

President

Director

Fig. 72. Certificate of the American Museum of Natural History, accompanying the award of its gold medal to Willi Hennig on January 23, 1975. Courtesy of Gerd Hennig, Tübingen (photo taken by M.S. on 30.10.2012).

CHAPTER SEVEN

THE "HENNIGIAN REVOLUTION"

Wikipedia defines a revolution as "a fundamental change in power or organizational structures that takes place in a relatively short period of time" (Doc. 171). If what Hennig presented caused indeed a revolution, then a "fundamental change" should be recognisable. As there is no general agreement on the meaning of "fundamental" in any concrete case, it is a matter of taste what one accepts as "fundamental". But just that there was and still is such a long and fervid argument about Hennig's systematics shows that there must be a fundamental disagreement between his approach and some earlier schools of science. When checking the methods and outcomes of traditional systematics and comparing them to the analyses done under the new paradigm, it becomes evident that there are indeed differences that could induce a feeling in traditional systematists of being threatened by the new style.

If it is not the practical method for revealing genealogical relationships, what then is the core of the "Hennigian revolution" of 1950 and 1966? Hennig introduced the necessity to systematics to make clear statements in the form "A is more closely related to B than either is to C" rather than put a taxon somewhere "in between" others or allegedly solve a taxonomic problem by opening a separate Linnean unit for a taxon in question. Moreover, he elaborated a method which required explicit presentation of supporting evidence rather than statements based purely on intuition or inexplicable experience. For the first time a method was at hand that made phylogenetics a scientific enterprise comparable to the branches of investigation which fall into Popper's concept of science (although there is still an ongoing debate on the question whether or not this applies to cladistics, i.e. the contemporary version of Hennigian phylogenetic systematics, see, e.g., Rieppel 2007a; Kluge 2009) (from Schmitt 2010a). Thus, one can firmly accept the expansion of the Hennigian method of systematics ("cladistics" for that matter) as a scientific revolution.

In his own review of the *Grundzüge*, Hennig elucidates the aspects of his book that he saw as most important (1952, no. 78a). The review emphasizes the relevance of systematics as a biological discipline that provides a general reference system for generalizations. This purpose can only be met, he argues, by a strictly phylogenetic system. He went on to criticize the conventional systems for mingling different concepts of 'relationship', especially the type of system "often denoted by the term 'network-relationship' [Netzverwandtschaft]". He does, however, not mention any newly introduced method. It is exclusively the concept (he called it the "theory") of systematics that he felt was noteworthy. From a present day's perspective it may seem unimpressive to invest such an amount of time and energy in the fight for an exclusively genealogical concept of 'relationship' in contrast to a concept based on 'similarity' of whatever kind (Figs. 73, 74). In the 1950s there was a strong opposition against such an innovation. In Germany, it was especially the school of Adolf Remane,

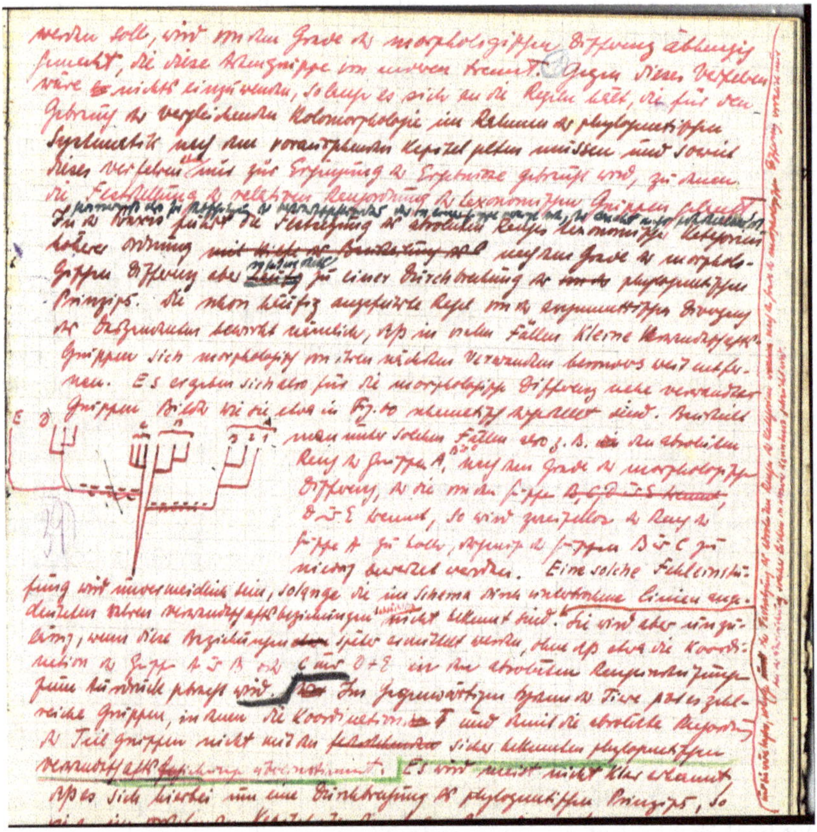

Fig. 73. Passage of the manuscript of the Grunduüge, corresponding to text on p. 208 of no. 76, the drawing became fig. 38. Courtesy of Wolfgang Hennig, Kranenburg.

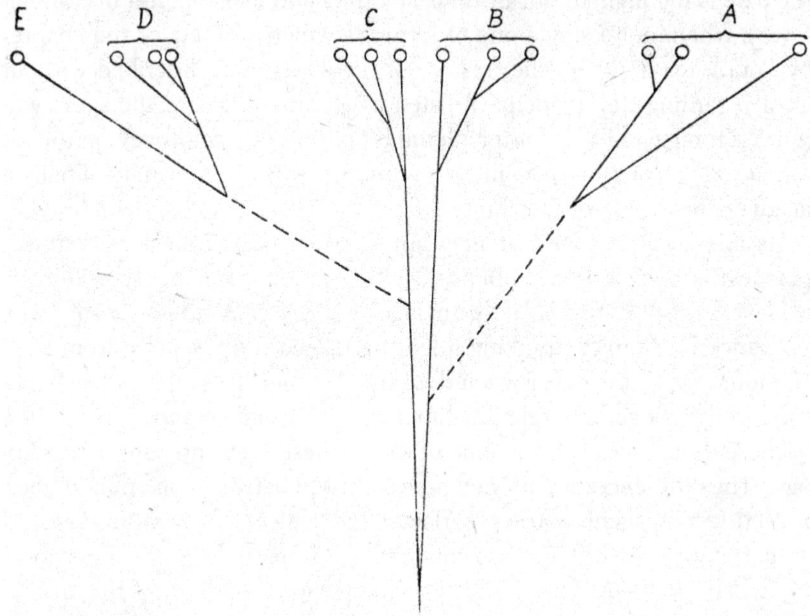

Fig. 74. Illustration of the crucial difference between similarity and (genealogical) relationship: A and B are more closely related to each other than any of them to C, although C and B are more similar to each other than any of them to any other taxon. Fig. 38 in the *Grundzüge* (1950, no. 76).

who insisted that taxa such as "Pongidae", "Reptilia" or "Apterygota" should be retained in a "natural" system. As a review of the first volumes of *Systematic Zoology* demonstrates (Hadži 1953), the state of systematics in North America was not significantly different.

Jahn (1992) mentions as one of Hennig's most outstanding contributions to systematics the new concept of "macrotaxonomy"—meaning a focus on relations between species and monophyletic groups of species—after about fifty years of prevalence of the relations between species and infraspecific populations. Likewise, Richter & Meier (1994: 212) stated that "Hennig was the person who redirected the interest of systematics to the study of supraspecific taxa after years of focusing on species and infraspecific taxa."

This means that in pre-Hennigian systematics a focus on the phylogenetic relations between supraspecific taxa was lacking, and that the ideas then current about how to form hypotheses about such relationships did not meet the scientific standards Hennig demanded. That this was indeed the case is highlighted, for instance, by the handbook chapter of Kühnelt (1962), where not a single one of Hennig's papers is cited, or the chapter by Franz (1943), where the "history of the animals" is described without even regarding the "principles" listed by Zimmermann in the same volume (Hennig cited these principles and built his own approach on them). Richter & Meier (1994) point to a similar picture in the major English-language textbooks of that time.

As already mentioned, it may appear surprising from a present-day perspective how much emphasis Hennig and those who accepted his achievements put on a "strictly phylogenetic" concept of relationship. The relevance of Hennig's thinking might be judged from a quote from J.S.L. Gilmour (1940, 461f.): "It is even doubtful whether the real significance of the term 'phylogenetic relationship' is yet fully understood. A resolution of these differences is surely one of the greatest needs of systematic biology." The differences mentioned here are those between 'taxonomic trees' reflecting 'phylogenetic trees' perfectly and 'logical classifications' (based on correlation or coherence of characters) not necessarily being 'phylogenetic'. This demand mirrors not just the personal opinion of the author, but a general understanding of that time, making clear that the publication of Hennig's *Grundzüge* marks a real innovation in biological systematics.

The actual revolution of systematics was certainly realised by the development of the practical method of polarising characters upon which modern cladistics is based. Hennig would most probably have appreciated this course of history. It is, however, highly unlikely that he would have accepted the complete abandonment of the traditional view of systematics as a discipline dealing with real organisms. Consequently, he would most probably have criticised the view of Platnick (1977) that the sister-group hypothesis "A is sister-group to B+C" is isomorphic to, for instance, the "phylogenetic tree" A→B→C (A is the ancestor of B, B is the ancestor of C). Following Hennig, each taxon, including the terminal taxa, has to be monophyletic. If they can be shown to be monophyletic, they must possess at least one autapomorphy and can, therefore, under no circumstances be ancestor to any other taxon of the above set (since an ancestor cannot have a character in an apomorph state as compared to its descendant). After all, the debate could only be sensible within a Hennigian framework if A, B, and C were species. There cannot be supraspecific ancestors. As Hull (1988: 137) pointed out "the only relation represented in cladograms and their isomorphic classifications are sister-group relations". As discussed above (chapter 5.1.), Hennig commented on the position of ancestors in cladograms only once, rather casually (in no. 125: 71; no. 161: 76).

Willi Hennig must definitively be credited for laying the foundations for the cladistic method of phylogenetic analysis although the unsatisfying elaboration of a practical procedure in his fundamental books of 1950 and 1966 leaves some doubt on whether it is justified to ascribe a new method as the turning point in the history of systematics to him. Even if his explicit descriptions of his practical method are theoretically insufficient and of only little practical use, he gave in his empirical papers on certain animals—mostly Diptera—valuable examples which could be used as models for further refinement of the method, last, but not least, in his general treatments of invertebrates (1957, 1959, 1969, see also Meier 2005). Certainly more important is the fact that he outlined a framework of clear terms and concepts that allows or even compels us to reduce subjectivity and arbitrariness in biological systematics. It is in this way that he made a crucial contribution to the conversion of systematics from an art or a handicraft to a legitimate branch of science. If this is regarded a revolution, then Willi Hennig was a revolutionary.

Box 7 Was Willi Hennig a Nazi?

Though Hennig opposed communist ideas, he was by no means a follower of National Socialist ("Nazi") politics. He was not a member of the NSDAP (Nationalsozialistische Deutsche Arbeiterpartei), as documented in the official dossier on him in the records of the State Museum of Natural History, Stuttgart (SMNS, Doc. 180), by his own entry in an official application form for a grant in 1936 (Doc. 173), and by the fact that there is no entry on him in the Berlin Document Center (Doc. 202). There is also no evidence of him holding Nazi attitudes in the written documents or verbal reports of his contemporaries, and even less evidence that he participated in any formally or morally condemnable activities during the Nazi period (and in general). Neither I nor Hennig's son Bernd have ever seen the Nazi salutation "Heil Hitler" used by Willi Hennig in his private correspondence. Even if a sender had written it, Hennig signed his reply letter using a conventional complimentary close (Doc. 078). Seemingly, he added this formula only to official letters to addressees in German institutions (Doc. 181, e.g. Doc. 032, Doc. 044, Doc. 182). Even Hennig's membership in the SA is no indication of any political conviction but—since he was not in addition a member of the NSDAP—a rather mild form of opportunism among academics (Doc. 191). One should not forget that Willi Hennig was twenty years old when he entered the SA, and he needed at that time the extension of a grant from the Studienstiftung des Deutschen Volkes, which had hastily turned into an integral part of the Nazi-machinery in 1933.

He was certainly not a member of the resistance, but it is absolutely clear anyway, that any accusation that Willi Hennig sympathized with Nazi ideology is pure invention. This view is also strongly supported by Rieppel (2011a).

Obviously, the singular source of such insinuations is a group of followers of Leon Croizat who extensively commented on the question "Was Hennig a Nazi?" on the internet discussion list Panbiog-L (Doc. 183). The origin of this denunciation is, according to Platnick & Nelson (1988: 415), Croizat's never substantiated personal anti-German attitude. When Gareth J. Nelson's wife asked Leon Croizat in the late 1970s why he considered Hennig a Nazi, he replied "Hennig was in the German army—he survived the war—he must have been a Nazi" (Doc. 185). Gareth Nelson was certain that Croizat's judgement of Hennig as a Nazi was based only on the surmise that Hennig, to have survived the war, must have been one of "them" (Doc. 186).

On Panbiog-L a posting by Robin Craw appeared in 2009: "Hennig was appointed to the Entomological Institute at Dahlem in 1937 under a grant from the DFG (German Research Organization). And who was the President of the DFG? Answer: SS-Brigadier General Ministeriel Direktor Professor Doktor Rudolf Mentzel. Yes, the fact is Hennig's early career was funded under the auspices of an SS-Brigadier General. What could be plainer than that?" (Doc. 184). Arguing that way implies that the recipient of a grant necessarily shares the political attitudes of the president of the funding organisation. To assume this in the absence of any further evidence is plainly unscientific. Much more important is, in my opinion, that the NS-regime governed Germany with totalitarian power. In 1937, there was simply no other science funding organisation in Germany. Hennig had no alternative. Demanding that he or any other person in a comparable situation should refrain from taking money from a public source (which meant inevitably: from a Nazi-governed institution) appears inexpressibly arrogant to me. The consequence would have been that Hennig would not have had the chance to become a professional scientist, and that he could not marry and found a family. To blame a man that he wanted to reach just these personal goals under difficult political and social conditions is nothing but mean.

Fig. 75. Part of Willi Hennig's office at the SMNS in Ludwigsburg, mid-1960s. Courtesy of Gerd Hennig, Tübingen.

CHAPTER EIGHT

WILLI HENNIG—A MAN OF ORDER

Willi Hennig can be seen as a rather unassertive (Schmitt 2010a) scout of the cladistic revolution that followed him. He prepared the soil for the great change and made the first step but did not show any intention to walk the path to the end.

There is an amusing parallel between Willi Hennig and Carl Linnaeus (also a firstborn, by the way): the latter is lauded for the introduction or at least the consequent use of binomial names for species, i.e. establishing the combination of a genus name and a "trivial" name. Indeed, he did not mention that when listing his major contribution to science in his autobiography (see Schmitt 2008). Thus, we possibly prize the two of them for achievements that they did not regard as their greatest themselves.

Even if little is left of Willi Hennig's original principles in modern cladistics, there is no doubt that he turned systematics from a craft or an art into a scientific enterprise that fits into the Popperian hypothetico-deductivistic model. He worked out a clear terminology and argumentation scheme (Wiley 1981; Richter & Meier 1994) and by this helped to eliminate or at least reduce arbitrariness in this field of science.

Clearly, Willi Hennig's primary aim was to establish a system that provides a proper and stable place for any species or supraspecific taxon. To reconstruct the genealogical relationships between the taxa studied was a necessary step on the way towards that goal, but it was certainly not Hennig's main interest. Although he mentioned functional aspects and adaptational considerations here and there, he was definitely not an evolutionary biologist. He was a man of order.

REFERENCES

1. *Unpublished Sources*

Where not stated otherwise, the documents are kept by the Hennig family (Mr. Gerd Hennig, Tübingen, Germany), and xerocopies are with the author.

KGBB: Klaus Günther, Berlin, bequest. Originals probably now at Staatsbibliothek Berlin ("Nachlaß 151"), seen and xerocopied by M.S. thanks to Waltraut Günther, in 1985.

SDEI: Senckenberg Entomologisches Institut, Müncheberg (Germany).

SMNS: Staatliches Museum für Naturkunde, Stuttgart (Germany).

Doc. 001. Handwritten autobiographical notes by Rudolf Hennig, original with Katharina Linke, xerocopy with M.S.

Doc. 002. http://www.loebaufoto.de/gschw10.htm, last visit April 2, 2012.

Doc. 003. Minutes of an interview of M.S. with Irma Hennig on August 7, 1995, in her home in Ludwigsburg-Pflugfelden.

Doc. 004. http://einestages.spiegel.de/static/topicalbumbackground/1278/zeit_ist_geld.html, last visit 19.04.2012.

Doc. 005. Minutes of an interview of Prof. Dr. Willi Xylander (Görlitz) with Mrs. R. Fabian on 15.10.1996, courtesy of W. Xylander.

Doc. 006. Letter of Wilhelm Meise to M.S., 12 pp., of 12.06.1998.

Doc. 007. Application for a grant from the Studienstiftung des Deutschen Volkes, without date, but written either before or shortly after entering university. Copy with Willi Hennig's son Dr. Bernd Hennig, Freiburg im Breisgau, seen by M.S. on 07.08.1995).

Doc. 008. Four pages of a diary of Willi Hennig 08.05.–03.06.1927.

Doc. 009. Composition test at school on "Die Stellung der Systematik in der Zoologie," 370 lines, 31 pp., handwritten, May 4, 1931. Xerocopy in the author's private archive, original with Dr. Bernd Hennig, Freiburg im Breisgau (Germany). This essay was posthumously published by Dieter Schlee in *Entomologica Germanica* 4: 193–199 (1978, see below no. 166).

Doc. 010. Abiturzeitung 1932A. Xerocopy through courtesy of W. Xylander with M.S.

Doc. 011. Willi Hennig's final school certificate (Schulentlassungszeugnis) dated 19.03.1927, seen by M.S. at Irma Hennig's home, 07.08.1995.

Doc. 012. Letter of Wilhelm Meise to M.S. of 04.07.1998, containing a transcript of a letter written by Fritz van Emden to Wilhelm Meise of 26.7.1948.

Doc. 013. Letter of Wilhelm Meise to M.S. of 17.06.1998.

Doc. 014. Minutes of an interview with Professor Helmut van Emden in London, 24.05.2012.

Doc. 015. Willi Hennig's record of study. Original with the family (Gerd Hennig, Tübingen), scan with M.S.

Doc. 016. http://www.uni-leipzig.de/unigeschichte/professorenkatalog//, last visit June 7, 2012.

Doc. 017. Notification of scholarship by Deutsches Studentenwerk—Studienstiftung des Deutschen Volkes to Willi Hennig, dated 19.05.1932.

Doc. 018. Letter of Fritz van Emden to Willi Hennig, dated 02.05.1932.

Doc. 019. Postcard of Fritz van Emden to Willi Hennig, dated 02.05.1932.

Doc. 020. "Rundschreiben an die Mitglieder und Vorsemester der Studienstiftung", dated 10.07.1933.

Doc. 021. Handwritten draft of a report to the Studienstiftung des Deutschen Volkes, 5 pp., undated but definitely written after July, 1933, and probably before October, 1933.

Doc. 022. Willi Hennig's duty card ("Pflichtenheft") for students.
Doc. 023. Letter of Fritz van Emden to Willi Hennig, dated 23.11.1932.
Doc. 024. Carbon copy of a letter of reference by Prof. Dr. G. Grimpe for Willi Hennig, dated 03.08.1933.
Doc. 025. http://de.wikipedia.org/wiki/Bekenntnis_der_Professoren_an_den_deutschen_Universitäten_und_Hochschulen_zu_Adolf_Hitler of 11.11.1933. Last visit 13.06.2012.
Doc. 026. Letter of Paul Buchner to Willi Hennig of 04.02.1958, written on the Island of Ischia. Original seen by M.S. at Irma Hennig's home on 04.12.1995.
Doc. 027. Letter of Prof. Koebe, dean of the philosophical faculty of the Leipzig university, to Willi Hennig, dated 7 June, 1935.
Doc. 028. Note from the dean's office to Willi Hennig, dated December 4, 1935.
Doc. 029. Letter of Klaus Günther to Willi Hennig, dated 25.05.1936. Original seen by M.S. at Irma Hennig's home on 04.12.1995.
Doc. 030. Permit to work scientifically at the Dresden State Museum of Zoology and Anthropology, dated 24.03.1936, signed by Dr. Hans Kummerlöwe.
Doc. 031. Letter of Karl Maria Heller to Willi Hennig, dated 31.12.1936.
Doc. 032.Letter of Willi Hennig to Walther Horn, dated 03.01.1937.
Doc. 033. Letter of Walther Horn to Willi Hennig, dated 05.01.1937, carbon copy at Senckenberg Deutsches Entomologisches Institut Müncheberg (SDEI), no. Hennig 39, photocopy with M.S.).
Doc. 034. http://de.wikipedia.org/wiki/Deutsche_W%C3%A4hrungsgeschichte. Last visit 22.06.2012.
Doc. 035. Letter of confirmation of 01.04.1937, signed by Walther Horn. Archive of the SDEI, no. Hennig/Horn 37, photocopy with M.S.
Doc. 036. Arbeitsbuch Nr. 40/1807487 Dr. Willi Hennig, first issued 07.10.1938.
Doc. 037. Personalbogen, signed on 11.03.1962 by Willi Hennig, to hand in at the Staatliches Museum für Naturkunde, Stuttgart, as a prerequisite for his employment there.
Doc. 038. Official letter of confirmation by Walther Horn, dated 14.04.1938.
Doc. 039. Letter of Carl Zimmer to Willi Hennig, dated 13.01.1937.
Doc. 040. E-mail from Gerd Hennig to M.S. of 08.08.2012.
Doc. 041. "Anmeldung bei der polizeilichen Meldebehörde" (registration with the police), issued 17.05.1939.
Doc. 042. Archive nos. Hennig 1, 4, 5, 6, 8, 9, 10, 15, and Sachtleben nos. 5, 16, 19, 26 in SDEI, photocopies with M.S.
Doc. 043. Letter of Max Beier to Willi Hennig, dated 01.03.1939.
Doc. 044. Letter of Willi Hennig to Otto Pesta, dated 05.03.1939.
Doc. 045. Letter of Otto Pesta to Willi Hennig, dated 08.03.1939.
Doc. 046. Certificate for the badge for casualties in Black, dated 22.05.1942.
Doc. 047. http://militaria-archiv.com/archive/index.php/t-17217.html, last visit 03.07.2012
Doc. 048. http://www.lexikon-der-wehrmacht.de/Gliederungen/Infanteriedivisionen/218ID-R.htm, last visit 03.07.2012.
Doc. 049. Army postal letters by Willi Hennig: 12 and one postcard from 1939, 27 (of 42 numbered) from 1941, 6 from 1943, 40 (of 42 numbered) from 1944 plus six letters written between April and June, 1944, and sent by civil mail from Berlin, 12 (of 18 numbered) from 1945 plus three written between January 5 and 11, and sent by civil mail from Berlin. Why and when the missing letters got lost is unclear, either they never arrived due to war circumstances, or they were not kept by Irma Hennig, for whatever reasons. All originals are kept by Dr. Bernd Hennig, Freiburg im Breisgau (Germany), scanned and printed copies kept by Mr. Gerd Hennig have been seen by M.S. in 2012.
Doc. 050. Pocket calendar for 1940 by "Stadtsparkasse—Stadtgirokasse Neusalza-Spremberg", ca. 7 × 10 cm, used by Willi Hennig, kept by Mr. Gerd Hennig, Tübingen (Germany), seen by M.S. in 2012.
Doc. 051. http://www.polunbi.de/bibliothek/1926-grimm-volk.html, last visit 10.07.2012.

Doc. 052. Letter from H. Sachtleben to W.H., dated 5.4.1941. Original at SDEI, Sachtleben 16, photocopy with M.S.

Doc. 053. Letter of Fritz Peus to W.H., dated 20.09.1941.

Doc. 054. Letter of Erich Martini to W.H., dated 05.10.1941.

Doc. 055. http://www.fachpublikationen.de/dokumente/01/05/index.html, last visit 08.08.2012.

Doc. 056. http://de.wikipedia.org/wiki/Schlacht_um_Cholm, last visit 07.08.2012.

Doc. 057. Page 22 of Willi Hennig's pay book. Photocopy with M.S.

Doc. 058. http://de.wikipedia.org/wiki/Medaille_Winterschlacht_im_Osten_1941/42, last visit 09.08.2012.

Doc. 059. Pocket calendar for 1942 with entries by Willi Hennig from January 1 to March 5, ca. 7 × 10 cm, used by Willi Hennig, kept by Mr. Gerd Hennig, Tübingen (Germany), seen by M.S. in 2012.

Doc. 060. Gesundheitsbuch of W.H., scans with M.S.

Doc. 061. Letter of Erich Martini to W.H., dated 16.02.1942.

Doc. 062. Letter of Hauptmann (name cannot be deciphered) to W.H., dated 24.02.1942.

Doc. 063. Letter of August Thienemann to W.H., dated 15.05.1942.

Doc. 064. Letter of August Thienemann to W.H., dated 28.05.1942.

Doc. 065. Letter of August Thienemann to W.H., dated 02.06.1942.

Doc. 066. Letter of W.H. to Hermann Eidmann, dated 27.10.1942.

Doc. 067. Letter of Hermann Eidmann to W.H., dated 06.11.1942.

Doc. 068. Army train tickets and seat reservation for a journey from Berlin to Saloniki and back, issued on 29.09.1943.

Doc. 069. Sammlung wehrrechtlicher Gutachten und Vorschriften, Heft 3, Bundesarchiv, Zentralnachweisstelle Kornelimünster. Printed as a manuscript 1965.

Doc. 070. Spiral-bound note-book, ca. A5, with itineraries and notes by Willi Hennig between September 30, 1943 and August 22, 1944.

Doc. 071. Grenzübertrittsausweis no. 50292 für Angehörige der deutschen Wehrmacht, issued November 15, 1943.

Doc. 072. http://www.archiv.sachsen.de/archive/leipzig/4180_3231303938.htm, last visit 09.08.2012.

Doc. 073. Letter of Rudolf Hennig to W.H., dated 14.01.1945. Original kept by Dr. Bernd Hennig, Freiburg im Breisgau (Germany), scanned and printed copy kept by Mr. Gerd Hennig, and seen by M.S. in 2012.

Doc. 074. Certificate for the military cross of merit, 2nd class with swords. Original kept by Mr. Gerd Hennig, photocopy with M.S.

Doc. 075. Note by Major J.A.W. Shearer on a report by W.H. in which Hennig's affiliation is given as "Abano Hospital + Centre", dated 28 May 1945.

Doc. 076. "Malaria and its control in liberated Italy from January to September 1944", 13 pp., document W0204/3010 at the National Archives, Kew, UK.

Doc. 077. Routine Order by Major-General C.A. Heydeman, serial no. 13, dated 11 April 1945, 4 pp., document W0204/6634 at the National Archives, Kew, UK.

Doc. 078. Letter of Dr. Bernd Hennig, Freiburg im Breisgau (Germany) to M.S., dated 04.01.2001.

Doc. 079. Original handwritten manuscript of "Grundzüge einer Theorie der Phylogenetischen Systematik", 170 pp., seen by M.S. on 07.08.1995 at Irma Hennig's home, now kept by Prof. Dr. Wolfgang Hennig, Kranenburg (Germany).

Doc. 080. Certificate of Discharge of Willi Hennig, issued October 16, 1945.

Doc. 081. Abrechnungen der Universitätsquästur Leipzig (financial acount of the university bursary) for winter-semester 1945/46, summer-semester 1946, and winter-semester 1946/47.

Doc. 082. Letter of the president of the Biologische Zentralanstalt für Land- und Forstwirtschaft to W.H., dated 26.02.1946.

Doc. 083. Letter of W.H. to the rector of the university of Leipzig, dated 25.03.1947.
Doc. 084. "Anmeldung bei der polizeilichen Meldebehörde" (registration with the police), issued 02.04.1947.
Doc. 085. Letter of W.H. to Irma Hennig, dated 21.07.1948. Original kept by Dr. Bernd Hennig, Freiburg im Breisgau (Germany), scanned and printed copy kept by Mr. Gerd Hennig, and seen by M.S. in 2012.
Doc. 086. Letter of W.H. to Irma Hennig, dated 29.07.1948, as Doc. 085.
Doc. 087. Letter of W.H. to Prof. Dr. C. Lehmann, director of the Institute of Zoology at the Brandenburgische Landeshochschule, dated 18.06.1950.
Doc. 088. Anmeldung für das polizeiliche Meldeamt, issued 07.10.1947.
Doc. 089. "Anmerkungen zu Personalfragen", written by Willi Hennig on 26.11.1966 on the occasion of an inquiry by the dean of the Faculty of Mathematics and Sciences of the Freie Universität Berlin, Prof. K.P. Grotemeyer, dated 15.11.1966 (as W.H. was put on rank two on the list of possible successors of Werner Ulrich). Original copy kept by the family, xerocopy with M.S.
Doc. 090. Letter of Irma Hennig to M.S., dated 19.08.1995.
Doc. 091. Letter of W.H. to the rector of the "Pädagogische Hochschule Potsdam" (until 1951: Brandenburgische Landeshochschule), dated 24.12.1953.
Doc. 092. Letter of E. Plachy to W.H., dated 04.02.1954.
Doc. 093. Letter of E. Plachy to W.H., dated 31.03.1955.
Doc. 094. Letter of Dr. Dyhrenfurth on behalf of E. Plachy to W.H., dated 01.07.1955.
Doc. 095. Letter of Paul Buchner to W.H., dated 03.06.1951.
Doc. 096. Questionnaire of the Netherlands for persons applying for a visa, with original signature and passport photo, dated 30.06.1951.
Doc. 097. Letter of Fritz van Emden to W.H., dated 18.02.1958. Archive of the SDEI, "Sachtleben 13", photocopy with M.S.
Doc. 098. Letter of W.H. to Fritz van Emden, dated 06.03.1958. Archive of the SDEI, "Sachtleben 14", photocopy with M.S.
Doc. 099. Letter of Hans Sachtleben to W.H., dated 27.06.1953.
Doc. 100. Single Round-Trip Interzonal Pass for Willi Hennig, issued 20.06.1953.
Doc. 101. Letter of Prof. Dr. Wolfgang Hennig to M.S., dated 09.01.1998.
Doc. 102. Letter of Rudolf Kleinert to W.H., dated 21.10.1957.
Doc. 103. Willi Hennig's dossier at the intelligence service of the GDR (Ministerium für Staatssicherheit, "Stasi"). 213 numbered pages plus two pages lacking pagination, all as official xercopies made by the Bundesbeauftragte für die Unterlagen des Staatssicherheitsdienstes der ehemaligen Deutschen Demokratischen Republik, plus 18 pages of pertaining correspondence, applied for and kept by Mr. Gerd Hennig (Tübingen, Germany).
Doc. 104. "Geschichte des DEI". http://www.dei-digital.de/book/export/html/38, last visit 31.08.2012.
Doc. 105. Letter of W.H. to Karl Strenzke, Wilhelmshaven, dated 05.03.1961.
Doc. 106. Minutes of a telephone conversation of Gerd Hennig (Tübingen, Germany) with M.S. on 03.09.2012.
Doc. 107. Letter of W.H. to the Central Administration of the Max-Planck-Gesellschaft, attn. Mr. Roeske, dated 28.07.1968. Original seen by M.S. at Irma Hennig's home on 04.12.1995.
Doc. 108. File card "Willi Hennig" in the biographical card file of SDEI (seen on 12.07.1995 at Eberswalde, Germany).
Doc. 109. Carbon copy of an application to the Deutsche Forschungsgemeinschaft by W.H. for a grant, dated 19.09.1961.
Doc. 110. Letter of W.H. to Dr. Meyl of Deutsche Forschungsgemeinschaft, Bad Godesberg, dated 14.12.1961.
Doc. 111. Letter of W.H. to Hans Sachtleben, dated 23.10.1961.
Doc. 112. Letter of Elmo Hardy to W.H., dated 31.01.1962.

Doc. 113. Letter of Judson Linsley Gressitt to W.H., dated 24.05.1962.
Doc. 114. Letter of Roland W. Force (director of the Bishop Museum) to W.H., dated 23.05.1962.
Doc. 115. Letter of W.H. to J.L. Gressitt, dated 07.06.1962.
Doc. 116. "1. Entwurf" (outline) of the configuration of a department of phylogenetic research by W.H., 5 pp., without date, but ca. 1962.
Doc. 117. Minutes of an interview of M.S. with Dr. Dieter Schlee (then SMNS) on 09.08.1995.
Doc. 118. Draft of a department's annual report by W.H., undated, SMNS archive.
Doc. 119. Letter of W.H. to Klaus Günther, dated 30.12.1963.
Doc. 120. Letter of W.H. to Klaus Günther, dated 30.01.1965.
Doc. 121. Letter of Klaus Günther to W.H., dated 05.03.1965, KGBB.
Doc. 122. Letter of W.H. to Klaus Günther, dated 30.12.1963, KGBB.
Doc. 123. Letter of Klaus Günther to W.H., dated 24.01.1964.
Doc. 124. Letter of W.H. to Joachim Illies, dated January, 1964. SMNS archive.
Doc. 125. Letter of W.H. to Graham C.D. Griffiths, dated 04.02.1964, SMNS archive.
Doc. 126. Letter of Graham C.D. Griffiths to W.H., dated 07.02.1964, SMNS archive.
Doc. 127. Letter of W.H. to Klaus Günther, dated 09.03.1965, KGBB.
Doc. 128. Letter of W.H. to Rainer Zangerl (Chicago), dated 01.03.1965, SMNS archive.
Doc. 129. Letter of W.H. to Gareth J. Nelson (New York, N.Y.), dated 27.09.1973, SMNS archive.
Doc. 130. Letter of W.H. to Klaus Günther, dated 06.06.1967.
Doc. 131. Letter of Klaus Günther to W.H., dated 11.06.1967.
Doc. 132. Minutes of an interview of M.S. with Dr. Gerd von Wahlert (then Ingersheim, Germany) on 08.08.1995, some information added after a telephone conversation on 21.09.1995.
Doc. 133. Letter of Walter Georg Kühne to W.H., dated 30.11.1965.
Doc. 134. Letter of W.H. to Klaus Günther, dated 05.12.1965.
Doc. 135 Letter of Klaus Günther to W.H., dated 22.12.1965.
Doc. 136. Minutes of a conversation of M.S. with Klaus Günther's widow Waltraut Günther, on 02.11.1985.
Doc. 137. Letter of Walter Georg Kühne to W.H., dated 05.11.1966.
Doc. 138. Letter of Klaus Günther to W.H., dated 04.11.1966.
Doc. 139. Letter of Klaus Günther to W.H., dated 09.12.1966.
Doc. 140. Letter of Walter Georg Kühne to W.H., dated 04.07.1967.
Doc. 141. Circular, sent to Peter Ax, Lars Brundin, Joachim Illies, Otto Kraus, Günther Osche, and Klaus Günther, and as carbon copy to W.H., by Walter Georg Kühne, dated 25.02.1969.
Doc. 142. Letter of support by Otto Kraus, dated 05.03.1969, carbon copy through Walter Georg Kühne to W.H.
Doc. 143. Letter of support by Günther Osche, dated 07.03.1969, carbon copy through Walter Georg Kühne to W.H.
Doc. 144. Letter of W.H. to Walter Georg Kühne, dated 14.03.1969.
Doc. 145. Letter of Adolf Butenandt, then president of the Max-Planck-Gesellschaft, to Walter Georg Kühne and Klaus Günther, dated 20.08.1969, carbon copy through W.G. Kühne to W.H.
Doc. 146. Letter of Klaus Günther to Irma Hennig, dated 04.12.1968.
Doc. 147. Letter of W.H. to Klaus Günther, dated 20.12.1968.
Doc. 148. Minutes of an interview of M.S. with Dr. Helmut Schmalfuß (then SMNS), on 06.12.1996.
Doc. 149. Minutes of an interview of M.S. with Drs. Gerhard & Erika Mickoleit (Tübingen) on 14.04.1997, with additions on 13.09.2012 (by telephone).
Doc. 150. Letter of Joachim Illies (Schlitz, Germany) to W.H., dated 25.03.1969, SMNS archive.

Doc. 151. Letter of W.H. to Joachim Illies, dated 26.03.1969, SMNS archive.
Doc. 152. University calendars for 1970–1976, library of the Eberhard-Karls-Universität, Tübingen.
Doc. 153. Minutes of an interview of M.S. with Ute Spahr (Pflugfelden), Willi Hennig's technician and gofer from 1966 to 1976 on 30.10.2000.
Doc. 154. Letter of W.H. to Günther Peters (Berlin), dated 28.02.1972, SMNS archive.
Doc. 155. Letter of Klaus Günther to W.H., dated 11.12.1973.
Doc. 156. Letter of Klaus Günther to W.H., dated 09.01.9174.
Doc. 157. Letter of Ernst Mayr (Museum of Comparative Zoology, Cambridge, MA, USA) to W.H., dated 11.09.1974, SMNS archive.
Doc. 158. Letter of Lars Zakarias Brundin to W.H., dated 18.04.1967.
Doc. 159. Letter of W.H. to Gareth J. Nelson (then: American Museum of Natural History, New York, NY, USA), dated 16.07.1973, SMNS archive.
Doc. 160. Letter of W.H. to Gareth J. Nelson (then: American Museum of Natural History, New York, NY, USA), dated 06.08.1973, SMNS archive.
Doc. 161. Letter of W.H. to Gareth J. Nelson (then: American Museum of Natural History, New York, NY, USA), dated 15.08.1973, SMNS archive.
Doc. 162. "Bericht über eine Studienreise nach Kanada und den USA" (report on a research trip to Canada and the USA), by Willi Hennig, SMNS archive.
Doc. 163. Willi Hennig's passport. Original kept by Gerd Hennig, Tübingen, photocopies with M.S.
Doc. 164. Travel documents for Willi and Irma Hennig, souvenirs, slides and notes, kept by Gerd Hennig, Tübingen.
Doc. 165. Medical report on Willi Hennig, dated 24.08.1976.
Doc. 166. Letter of W.H. to Günther Osche (Freiburg im Breisgau), dated 20.08.1975, SMNS archive.
Doc. 167. E-mail from Gerd Hennig (Tübingen) to M.S., sent 18.09.2012.
Doc. 168. E-mail from Dr. Bernd Hennig (Freiburg im Breisgau) to M.S., sent 18.09.2012.
Doc. 169. Letter of W.H. to Lars Brundin (Stockholm), dated 25.01.1971, SMNS archive.
Doc. 170. Letter of Prof. Dr. Klaus Sander (Freiburg im Breisgau) to W.H., dated 04.11.1964, reply from W.H. dated 09.11.1964, SMNS archive.
Doc. 171. http://en.wikipedia.org/wiki/Revolution, last visit 21.09.2012.
Doc. 172. E-mail from Prof. Dr. Wolfgang Hennig (Kranenburg, Germany) to M.S., on 22.09.2012.
Doc. 173. Dossier on Willi Hennig of the former archive of the Deutsche Forschungsgemeinschaft, now at Bundesarchiv Berlin-Lichterfelde, signature R 73/11600.
Doc. 174. Official information of the Deutsche Dienststelle für die Benachrichtigung der nächsten Angehörigen von Gefallenen der ehemaligen deutschen Wehrmacht (Berlin) to Mr. Gerd Hennig (Tübingen), dated 11.09.2012, photocopy with M.S.
Doc. 175. E-mail from Prof. Dr. Wolfgang Hennig (Kranenburg, Germany) to M.S., on 27.09.2012.
Doc. 176. Letter of the Bundesbeauftragte für die Unterlagen des Staatssicherheitsdienstes der ehemaligen Deutschen Demokratischen Republik to Gerd Hennig (Tübingen), dated 15.07.2008.
Doc. 177. Tape record of a lecture presented by Klaus Günther to the symposium on methods of phylogenetics, held on 13.02.1970, plus the following discussion. Taken by Prof. Dr. Erwin Tretzel, Kaiserslautern (Germany), copy with M.S.
Doc. 178. Letter of W.H. to Prof. Fritz Peus, Berlin, dated 04.02.1966, SMNS archive.
Doc. 179. Letter of W.H. to the Deutsche Forschungsgemeinschaft, dated 11.03.1968.
Doc. 180. "Bericht der Abt. für Phylogenetik 3.1962, Anlage 2" in Willi Hennig's dossier at Staatliches Museum für Naturkunde, Stuttgart, SMNS archive.
Doc. 181. E-message from Niels Peder Kristensen, Copenhagen (Denmark) to M.S. of 17.07.2010.
Doc. 182. Scanned image of two letters dated 12.10. and 21.10.1940, by Willi Hennig to O. Schröder, at that time curator at the Zoological Museum of Kiel. Originals at the Zoological Museum, Copenhagen (courtesy of N.P. Kristensen).

Doc. 183. Digest of the discussion thread "Der Biologe, German neo-Darwinism and Hennig" of April 2, 2009 (http://groups.google.com/group/panbiog/t/05e0fcb97dcf655e?hl=en).
Doc. 184. E-mail from Gareth J. Nelson (Melbourne, Australia) to M.S. of 31.03.2009.
Doc. 185. E-mail from Gareth J. Nelson (Melbourne, Australia) to Olivier Rieppel (Chicago IL, USA) of 08.08.2011 and forwarded to M.S on 09.08.2011.
Doc. 186. E-mail from Gareth J. Nelson (Melbourne, Australia) to M.S. of 30.12.1998.
Doc. 187. Bock, W.J. & Hennig, W. 1974. Methods of phylogenetic analysis. 59 pp., typewritten manuscript, never published, courtesy of Gerd von Wahlert (then Ingersheim, Germany).
Doc. 188. Bock, W.J. & Hennig, W. 1974. Methods of phylogenetic analysis—an abstract. 13 pp., typewritten manuscript, never published, courtesy of Gerd von Wahlert (then Ingersheim, Germany).
Doc. 189. Letter of W.H. to Dr. Gareth J. Nelson (then: American Museum of Natural History, New York, NY, USA), dated 11.06.1976, SMNS archive.
Doc. 190. E-Mail of Quentin D. Wheeler (Arizona State University, Tempe, AZ, USA) to Malcolm Scoble (Natural History Museum. London, UK) and forwarded to M.S. of 12.09.2012.
Doc. 191. E-Mail of Prof. Dr. Götz Aly (Berlin) to M.S. of 03.10.2012.
Doc. 192. Letter of Gareth J. Nelson to W.H., dated 30.04.1975, SMNS archive.
Doc. 193. Letter of W.H. to Gareth J. Nelson (then: American Museum of Natural History, New York, NY, USA), dated 02.06.1975, SMNS archive.
Doc. 194. E-mail from Prof. Dr. Wolfgang Hennig (Kranenburg, Germany) to M.S. of 29.09.2012.
Doc. 195. E-mail from Prof. Dr. Wolfgang Hennig (Kranenburg, Germany) to M.S. of 04.10.2012.
Doc. 196. E-mail from Mr. Gerd Hennig (Tübingen, Germany, to M.S. of 06.10.2012.
Doc. 197. Letter of W.H. to Gareth J. Nelson (then: American Museum of Natural History, New York, NY, USA), dated 27.09.2973, SMNS archive.
Doc. 198. Letter of W.H. to Klaus Günther, dated 23.01.1966.
Doc. 199. E-Mail from Dr. Michael Buddrus (Institut für Zeitgeschichte, Berlin) to M.S. of 05.10.2012.
Doc. 200. Letter of W.H. to Klaus Günther, dated 10.08.1964, KGBB, xerocopy with M.S.
Doc. 201. Letter of Dr. Bernd Hennig, Freiburg im Breisgau (Germany) to M.S. of 24.10.2012.
Doc. 202. E-mail from Ms. Undine Beier (Bundesarchiv, Berlin), to M.S. of 26.10.2012.
Doc. 203. E-mail from Mr. Gerd Hennig (Tübingen, Germany, to M.S. of 06.11.2012.
Doc. 204. Card files on W.H. of the Reichsforschungsrat ("Hel 2/08"), at Bundesarchiv Berlin (formerly Berlin Document Center), R 3-2012/D-10023, xerocopies with M.S.
Doc. 205. Comment of Prof. Dr. Wolfgang Hennig, via e-mail to M.S. on 09.11.2012.
Doc. 206. Personnel review by the ministry of agrigulture and forestry of the GDR of 12.02.1951. Bundesarchiv Berlin DR/3/B, 11422/a, xerocopy with M.S.

2. *Published Sources*

2.1. *Willi Hennig's publications (numbering according to the list of Willi Hennig's publication, compiled by Dr. Bernd Hennig and published as Anonymus 1978; numbers in square brackets added by M.S.)*

[168] Hennig, W. 1931. Einiges über die Insekten des Landesschulgebietes. Mitteilungen aus der Landesschule Dresden, Heft 8: 1–6.
1. Meise, W. & Hennig, W. 1932. Die Schlangengattung *Dendrophis*. Zoologischer Anzeiger 99: 273–297.
2. Hennig, W. 1934. Zur Kenntnis der Kopulationsorgane der Tyliden (Mikropeziden, Dipt. Acalypt.). Zoologischer Anzeiger 107: 67–76.

3. Hennig, W. 1934. Über Bau und Verwandtschaft der Kerguelenfliege *Calycopteryx moseleyi* Eat. Zoologischer Anzeiger 108: 196–201.
4. Hennig, W. 1934/1935. Revision der Tyliden (Dipt., Acalypt.). I. Teil: Die Taeniapterinae Amerikas. Stettiner entomologische Zeitung 95: 65–108, 294–330 (1934), 96: 27–67 (1935).
5. Meise, W. & Hennig, W. 1935. Zur Kenntnis von *Dendrophis* und *Chrysopelea*. Ein Beitrag zur systematischen Bewertung der Opisthoglypha. Zoologischer Anzeiger 109: 138–150.
6. Hennig, W. 1935. Thalassobionte und thalassophile Diptera Nematocera. Pp. 85–102 in Grimpe, G. (ed.) Die Tierwelt der Nord- und Ostsee, Teil XI, e3, Lieferung 27. Akademische Verlagsgesellschaft, Leipzig.
7. Hennig, W. 1935. Der Filterappart im Pharynx der Cyclorrhaphenlarven und die biologische Bedeutung der Madenform. Zoologischer Anzeiger 111: 131–139.
8. Hennig, W. 1935/1936. Revision der Tyliden (Dipt., Acalypt.). II. Teil: Die außeramerikanischen Taeniapterinae, die Trepidariinae und Tylinae. Allgemeines über die Tyliden. (Zugleich ein Beitrag zu den Ergebnissen der Sunda-Expedition Rensch, 1927). Konowia 14: 68–92, 192–216, 289–310 (1935), 15: 129–144, 201–239 (1936).
9. Hennig, W. 1936. Revision der Gattung *Draco* (Agamidae). Temminckia 1: 153–220.
10. Hennig, W. 1936. Beiträge zur Kenntnis des Kopulationsapparates der cyclorrhaphen Dipteren. Zeitschrift für Morphologie und Ökologie der Tiere 31: 328–370.
11. Hennig, W. 1936. Der männliche Kopulationsapparat der Dipteren III—Psychodidae und Asilidae. Zoologischer Anzeiger 114: 177–186.
12. Hennig, W. 1936. Systematisch-tiergeographische Beiträge zur Kenntnis der Tethiniden (Dipt. Acalypt.). Entomologische Rundschau 54: 136–140.
13. Hennig, W. 1936. Beiträge zur Kenntnis des Kopulationsapparates und der Systematik der Tanypeziden (Dipt., Acalypt.). Deutsche Entomologische Zeitschrift 1936: 27–38.
14. Hennig, W. 1936. Über einige Gesetzmäßigkeiten der geographischen Variation in der Reptiliengattung *Draco* L.: "parallele" und "konvergente" Rassenbildung. Biologisches Zentralblatt 56: 549–559.
15. Hennig, W. 1936. Beiträge zur Systematik und Tiergeographie der Pyrgotiden (Diptera). Arbeiten über morphologische und taxonomische Entomologie aus Berlin-Dahlem 3: 243–256.
16. Hennig, W. 1936. Beziehungen zwischen geographischer Verbreitung und systematischer Gliederung bei einigen Dipterenfamilien: ein Beitrag zum Problem der Gliederung systematischer Kategorien höherer Ordnung. Zoologischer Anzeiger 116: 161–175.
17. Hennig, W. 1937. Übersicht über die Arten der Neriiden und über die Zoogeographie dieser Acalyptratengruppe (Diptera). Stettiner Entomologische Zeitung 98: 240–280.
18. Hennig, W. 1937. Nachträge zur "Revision der Tyliden". Stettiner entomologische Zeitung 98: 46–50.
19. Hennig, W. 1937. Tylidae (Micropezidae, Diptera) aus der Sammlung des British Museum of Natural History. Nachträge zur "Revision der Tyliden"—II. Annals and Magazine of Natural History Series 10, 19: 521–523.
20. Hennig, W. 1937. Die morphologische Deutung des männlichen Kopulationsapparates der Gattung *Glossina*. Zeitschrift für Parasitenkunde 9(3): 345–350.
21. Hennig, W. 1937. Neue Pyrgotiden aus dem Deutschen Entomologischen Institut. Arbeiten über morphologische und taxonomische Entomologie 4(3): 207–209.
22. Hennig, W. 1937. Tanypezidae. 6 pp. in Lindner, E. (ed.) Die Fliegen der paläarktischen Region, Teil 44, Lieferung 106. E. Schweizerbart'sche Verlagsbuchhandlung, Stuttgart.
23. Hennig, W. 1937. Coelopidae. 39 pp. in Lindner, E. (ed.) Die Fliegen der paläarktischen Region, Teil 52, Lieferung 106. E. Schweizerbart'sche Verlagsbuchhandlung, Stuttgart.
24. Hennig, W. 1937. Milichiidae et Carnidae. 91 pp. in Lindner, E. (ed.) Die Fliegen der paläarktischen Region, Teil 60a, Lieferung 115. E. Schweizerbart'sche Verlagsbuchhandlung, Stuttgart.

25. Hennig, W. 1937/1938. Beiträge zur Systematik der Richardiiden (Dipt.). Revista de Entomologia 7(1): 21–34, (2–3): 306–316, (4): 484–486, 1938 8(1–2): 111–122.
26. Hennig, W. 1938. Neue Übersicht über die Arten der Gattung *Glyphodera* End. (Dipt. Acalypt., Tylidae). Annals and Magazine of Natural History series 11, 1: 308–312.
27. Hennig, W. 1938. Neue Beiträge zur Systematik der Richardiiden und Tyliden (Diptera, Acalyptrata). Arbeiten über morphologische und taxonomische Entomologie aus Berlin-Dahlem 5(1): 8–15.
28. Hennig, W. 1938. Beiträge zur Kenntnis der Clusiiden und ihres Kopulationsapparates (Diptera: Acalyptrata). Pp. 121–138 in Encyclopédie entomologique: Recueil d'études biologiques et systématiques sur les diptères du globe. Série B, Mémoires et Notes. 2, Diptera. Lechevalier, Paris.
29. Hennig, W. 1938. Zur Frage der verwandtschaftlichen Stellung von *Braula coeca* Nitzsch. Arbeiten über morphologische und taxonomische Entomologie aus Berlin-Dahlem 5(2): 164–174 + 1 plate.
30. Hennig, W. 1938. Übersicht über die Larven der wichtigsten deutschen Chrysomelinen (Coleoptera). Arbeiten über physiologische und angewandte Entomologie aus Berlin-Dahlem 5: 85–136, 2 plates.
31. Hennig, W. 1938. Die Gattung *Rhachicerus* und ihre Verwandten im Baltischen Bernstein (Diptera). Zoologischer Anzeiger 123: 33–41.
32. Hennig, W. 1938. Zur Nomenklatur pflanzenschädlicher Anthomyiiden. Arbeiten über physiologische und angewandte Entomologie aus Berlin-Dahlem 5: 279–284.
33. Hennig, W. 1938. Tyliden aus Japan. Insecta Matsumurana 13: 1–14.
34. Hennig, W. 1938. Beiträge zur Kenntnis des Kopulationsapparates und der Systematik der Acalyptraten I. Chamaemyiidae und Odiniidae (Diptera). Arbeiten über morphologische und taxonomische Entomologie aus Berlin-Dahlem 5(3): 201–213, 1 plate.
35. Hennig, W. 1938. Odiniidae. 11 pp. in Lindner, E. (ed.) Die Fliegen der paläarktischen Region, Teil 60b, Lieferung 122. E. Schweizerbart'sche Verlagsbuchhandlung, Stuttgart.
36. Hennig, W. 1938. Braulidae. 14 pp. in Lindner, E. (ed.) Die Fliegen der paläarktischen Region, Teil 60c, Lieferung 122. E. Schweizerbart'sche Verlagsbuchhandlung, Stuttgart.
37. Hennig, W. 1939. Beiträge zur Kenntnis des Kopulationsapparates und der Systematik der Acalyptraten II. Tethinidae, Milichiidae, Anthomyzidae und Opomyzidae (Diptera). Arbeiten über morphologische und taxonomische Entomologie aus Berlin-Dahlem 6(2): 81–94, 1 plate.
38. Hennig, W. 1939. Über einen Floh aus der Bernsteinsammlung des Herrn Scheele (Aphaniptera: Ctenopsyllidae). Arbeiten über morphologische und taxonomische Entomologie aus Berlin-Dahlem 6: 330–332.
39. Hennig, W. 1939. Über Namen und Artenzahl der deutschen "Vogelblutfliegen". Arbeiten über physiologische und angewandte Entomologie aus Berlin-Dahlem 6: 359–364.
40. Hennig, W. 1939. Otitidae. 78 pp. in Lindner, E. (ed.) Die Fliegen der paläarktischen Region, Teil 46/47, Lieferung 126/128. E. Schweizerbart'sche Verlagsbuchhandlung, Stuttgart.
41. Hennig, W. 1939. Nachtrag zu meiner Bearbeitung der Otitidae in "Lindner, Die Fliegen der paläarktischen Region (Teil 46/47)". Arbeiten über morphologische und taxonomische Entomologie aus Berlin-Dahlem 6(3): 266–267.
42. Hennig, W. 1940. Kritische Übersicht über die verwandtschaftliche Stellung der bisher als "Phytalmiidae" zusammengefaßten Gattungen (Diptera). Arbeiten über morphologische und taxonomische Entomologie aus Berlin-Dahlem 7: 58–64.
43. Hennig, W. 1940. Über die systematische Stellung von *Palaetimia hoesti* Meun. Arbeiten über morphologische und taxonomische Entomologie aus Berlin-Dahlem 7: 89–91.
44. Hennig, W. 1940. Orientalische und papuanische Bibioniden im Deutschen Entomologischen Institut (Diptera). Arbeiten über morphologische und taxonomische Entomologie aus Berlin-Dahlem 7: 251–255.

45. Hennig, W. 1940. Außereuropäische Psiliden und Platystomiden im Deutschen Entomologischen Institut (Diptera). Arbeiten über morphologische und taxonomische Entomologie aus Berlin-Dahlem 7: 304–318.
46. Hennig, W. 1940. Ulidiidae. 34 pp., 6 plates, in Lindner, E. (ed.) Die Fliegen der paläarktischen Region, Teil 45, Lieferung 133. E. Schweizerbart'sche Verlagsbuchhandlung, Stuttgart.
47. Hennig, W. 1941. Verzeichnis der Dipteren von Formosa. Entomologische Beihefte aus Berlin-Dahlem 8: IV + 239 pp.
48. Hennig, W. 1941. Werden alle "Möhrenfliegen-Schäden" durch *Chamaepsila rosae* F. verursacht? Arbeiten über physiologische und angewandte Entomologie aus Berlin-Dahlem 8: 36–38.
49. Hennig, W. 1941. Dipteren von den kleinen Sunda-Inseln. Aus der Ausbeute der Sunda-Expedition Rensch. I. Einleitung und Bearbeitung der Familien Sciomyzidae, Tylidae, Lonchaeidae, Pyrgotidae und Platystomidae. Arbeiten über morphologische und taxonomische Entomologie aus Berlin-Dahlem 8(1): 16–23.
50. Hennig, W. 1941. Beiträge zur Kenntnis des Kopulationsapparates und der Systematik der Acalyptraten III. Pallopteridae, Thyreophoridae, Diopsidae, Pseudopomyza, Pseudodinia (Diptera). Arbeiten über morphologische und taxonomische Entomologie aus Berlin-Dahlem 8(1): 54–65 + 1 plate.
51. Hennig, W. 1941. *Seioptera*, eine für die taxonomische Methodik interessante Dipterengattung (Diptera: Otitidae). Arbeiten über morphologische und taxonomische Entomologie aus Berlin-Dahlem 8(1): 73–76.
52. Hennig, W. 1941. *Formicosepsis* de Meijere und *Cypselosoma* Hendel, zwei Gattungen der Tyliden—Nachtrag zur Revision der Tyliden (Diptera). Stettiner Entomologische Zeitung 102: 129–131.
53. Hennig, W. 1941. Dipteren von den kleinen Sunda-Inseln. Aus der Ausbeute der Sunda-Expedition Rensch. III. Piophilidae und Sepsidae. Arbeiten über morphologische und taxonomische Entomologie aus Berlin-Dahlem 8(3): 145–149.
54. Hennig, W. 1941. Neues über parasitische Dipteren (Cecidomyiidae und Chloropidae). Arbeiten über morphologische und taxonomische Entomologie aus Berlin-Dahlem 8: 196–200.
55. Hennig, W. 1941. Die Verwandtschaftsbeziehungen der Pupiparen und die Morphologie der Sternalregion des Thorax der Dipteren. Arbeiten über morphologische und taxonomische Entomologie aus Berlin-Dahlem 8: 231–249, 4 plates.
56. Hennig, W. 1941. Megamerinidae. 4 pp. in Lindner, E. (ed.) Die Fliegen der paläarktischen Region, Teil 39b, Lieferung 143. E. Schweizerbart'sche Verlagsbuchhandlung, Stuttgart.
57. Hennig, W. 1941. Diopsidae. 8 pp. in Lindner, E. (ed.) Die Fliegen der paläarktischen Region, Teil 39c, Lieferung 143. E. Schweizerbart'sche Verlagsbuchhandlung, Stuttgart.
58. Hennig, W. 1941. Psilidae. 38 pp. + 4 plates in Lindner, E. (ed.) Die Fliegen der paläarktischen Region, Teil 41, Lieferung 140. E. Schweizerbart'sche Verlagsbuchhandlung, Stuttgart.
59. von Blumenthal, R. & Hennig, W. 1942. Krankheitsübertragende Gliederfüßler (außer Fiebermücken) im Nahen Osten. 2 pp. in Zeiss, H. (ed.) Seuchen-Atlas, Nr. II/3 (1./2. Lieferung). Justus Perthes, Gotha.
60. Hennig, W. 1943. Verbreitung der Sandmücken (Gattung *Phlebotomus*) in Europa. 2 pp., in Zeiss, H. (ed.) Seuchen-Atlas, Nr. I/8 (3. Lieferung). Justus Perthes, Gotha.
61. von Blumenthal, R. & Hennig, W. 1943. Sandmücken und Zecken als Krankheitsüberträger im Transkaspischen Raum. 4 pp., in Zeiss, H. (ed.) Seuchen-Atlas, Nr. III/2 (3. Lieferung). Justus Perthes, Gotha.
62. Hennig, W. 1943. Übersicht über die bisher bekannten Metamorphosestadien der Ephydriden, mit Neubeschreibungen nach dem Material der Deutschen Limnologischen Sunda-Expedition (Diptera: Ephydridae). Arbeiten über morphologische und taxonomische Entomologie aus Berlin-Dahlem 10: 105–136.

63. Hennig, W. 1943. Ein Beitrag zum Problem der "Beziehungen zwischen Larven- und Imaginalsystematik". Arbeiten über morphologische und taxonomische Entomologie aus Berlin-Dahlem 10: 138–144.
64. Hennig, W. 1943. Die Larve von *Haplegis nigritarsis* Duda (Diptera: Chloropidae). Arbeiten über physiologische und angewandte Entomologie aus Berlin-Dahlem 10: 117–123.
65. Hennig, W. 1943. Einiges über die Metamorphose von *Megamerina dolium* Fabr. (= *loxocerina* Fall.) (Diptera: Acalyptrata: Megamerinidae). Arbeiten über morphologische und taxonomische Entomologie aus Berlin-Dahlem 10: 205–208.
66. Hennig, W. 1943. Piophilidae. 52 pp. in Lindner, E. (ed.) Die Fliegen der paläarktischen Region, Teil 40, Lieferung 151. E. Schweizerbart'sche Verlagsbuchhandlung, Stuttgart.
67. Hennig, W. 1945. Platystomidae. 56 pp. + 3 plates in Lindner, E. (ed.) Die Fliegen der paläarktischen Region, Teil 48, Lieferung 155. E. Schweizerbart'sche Verlagsbuchhandlung, Stuttgart.
68. Hennig, W. 1947. Probleme der biologischen Systematik. Forschungen und Fortschritte 21/23: 276–279.
69. Hennig, W. 1948. Eine neue Art der Gattung *Meoneura* aus den Niederen Tauern (Diptera, Milichiidae). Zeitschrift der Wiener Entomologischen Gesellschaft 33: 138–139.
70. Hennig, W. 1948. Beiträge zur Kenntnis des Kopulationsapparates und der Systematik der Acalyptraten IV "Lonchaeidae" und "Lauxaniidae". Acta zoologica Lilloana 6: 333–429.
71. Hennig, W. 1948. Über einige verkannte Dipteren-Gattungen. Acta zoologica Lilloana 6: 169–170.
72. Hennig, W. 1948–1952. Die Larvenformen der Dipteren. Eine Übersicht über die bisher bekannten Jugendstadien der zweiflügeligen Insekten. 1. Teil 185 pp., 1948, 2. Teil 458 pp., 1950, 3. Teil 628 pp, 1952. Akademie-Verlag, Berlin (unmodified reprint in 1968).
73. Hennig, W. 1949. Zur Klärung einiger Begriffe der phylogenetischen Systematik. Forschungen und Fortschritte 25: 137–139.
74. Hennig, W. 1949. Sepsidae. 91 pp. + 10 plates in Lindner, E. (ed.) Die Fliegen der paläarktischen Region, Teil 39a, Lieferung 157/159. E. Schweizerbart'sche Verlagsbuchhandlung, Stuttgart.
75. Hennig, W. 1950. Entomologische Beobachtungen an kleinen Wirbeltierleichen. Zeitschrift für hygienische Zoologie 38: 33–88.
76. Hennig, W. 1950. Grundzüge einer Theorie der phylogenetischen Systematik. 370 pp., Deutscher Zentralverlag, Berlin.
76a. Hennig, W. 1953. On the nature of systematics. Passages of pp. 4 and 10 from no. 76, translated by George C. Steyskal. Systematic Zoology 2(1): 41.
77. Hennig, W. 1951. Neue Acalyptraten aus Europa und Südafrika (Diptera: Piophilidae, Helomyzidae, Lauxaniidae). Beiträge zur Entomologie 1(1): 70–76.
78. Hennig, W. 1952. Dipteren von den Kleinen Sunda-Inseln. Aus der Ausbeute der Sunda-Expedition Rensch. IV. Fam. Muscidae. Beiträge zur Entomologie 2(1): 55–93.
78a. Hennig, W. 1952. Autorreferat: Grundzüge einer Theorie der phylogenetischen Systematik. Beiträge zur Entomologie 2: 339–331.
79. Hennig, W. 1952. Bemerkenswerte neue Acalyptraten in der Sammlung des Deutschen Entomologischen Institutes (Diptera: Acalyptrata). Beiträge zur Entomologie 2(6): 604–618.
80. Hennig, W.; Bollmann, H. & Machatschke, J. 1953. Kritische Bemerkungen zum phylogenetischen System der Insekten. Beiträge zur Entomologie 3 (Sonderheft Festschrift Sachtleben): 1–85.
81. Hennig, W. 1953. Diptera, Zweiflügler. Pp. 1–166 in Blunck, H. (ed.) Handbuch der Pflanzenkrankheiten, founded by Paul Sorauer, vol. 5/2, 5th ed., Lieferung 1. Paul Parey, Berlin – Hamburg.

82. Hennig, W. 1953. Übersicht über die europäischen Arten der Gattung *Chiastochaeta* (Diptera: Muscidae). Beiträge zur Entomologie 3(6): 655–668.
83. Hennig, W. 1953. Die Gattungen *Delia* und *Paregle* in Australien (Diptera: Muscidae). Beiträge zur Entomologie 3(6): 668–671.
84. Hennig, W. 1954. Flügelgeäder und System der Dipteren unter Berücksichtigung der aus dem Mesozoikum beschriebenen Fossilien. Beiträge zur Entomologie 4: 245–388.
85. Hennig, W. 1954. Die Kleewurzelhalsfliege: *Psila gracilis* Meig., nicht *Psila atra* Meig. (Diptera: Psilidae). Beiträge zur Entomologie 4: 544–545.
86. Hennig, W. 1954. Piophilidae und Sepsidae aus Ost-Afrika (Diptera). Beiträge zur Entomologie 4: 641–643.
87. Hennig, W. 1955. Das Flügelgeäder der Gattung *Allactoneura*. Eine Berichtigung (Diptera: Fungivoroidea). Beiträge zur Entomologie 5(1/2): 127–128.
88. Hennig, W. 1955. Bemerkungen zur Synonymie einiger Gattungen der niederen Brachycera (Diptera: Rhagionidae und Stratiomyiidae). Beiträge zur Entomologie 5(3/4): 426–428.
89. Hennig, W. 1955. Meinungsverschiedenheiten über das System der niederen Insekten. Zoologischer Anzeiger 155: 21–30.
90. Hennig, W. 1955. Los Insectos de las Islas Juan Fernández 16. Phryneidae, Helomyzidae, Lonchaeidae, Piophilidae, Anthomyzidae und Muscidae (Diptera). Revista Chilena de Entomología 4: 21–34.
91. Hennig, W. 1956. Beitrag zur Kenntnis der Milichiiden-Larven (Diptera: Milichiidae). Beiträge zur Entomologie 6(1/2): 138–145.
92. Hennig, W. 1956. Neue neotropische Acalyptrata aus dem Deutschen Entomologischen Institut (Diptera: Acalyptrata). Beiträge zur Entomologie 6(1/2): 146–154.
93. van Emden, F. & Hennig, W. 1956. Diptera. Pp. 111–122 in Tuxen, S.L. (ed.) Taxonomist's Glossary of Genitalia in Insects. Munksgaard, Copenhagen (2nd ed. 1970).
94. Hennig, W. 1957. Los Insectos de las Islas Juan Fernández 34. Therevidae, Helomyzidae, Piophilidae, Milichiidae und Muscidae (Diptera) (Nachtrag). Revista Chilena de Entomología 5: 409–412.
95. Hennig, W. 1957. Systematik und Phylogenese. Pp. 50–71 in: Hannemann, H.-J. (ed.) Bericht über die Hundertjahrfeier der Deutschen Entomologischen Gesellschaft Berlin 1956. Akademie-Verlag, Berlin.
96. Hennig, W. 1957. Taschenbuch der Zoologie Heft 2, Wirbellose I, ausgenommen Gliedertiere, 1st ed., VI + 147 pp., Georg Thieme; Leipzig.
97. Piekarski, G.; Hennig, W. & Sibbing, B. 1957: Die geographische Verbreitung der Leishmaniasen und der Phlebotomen als ihrer Überträger in Mittel- und Südamerika (1911–1955). Text III/79–81, maps 100–120/VII in Rodenwaldt, E. (ed.) Weltseuchenatlas/World Atlas of Epidemic Diseases. Falk, Hamburg.
98. Hennig, W. 1958. Die Familien der Diptera Schizophora und ihre phylogenetischen Verwandtschaftsbeziehungen. Beiträge zur Entomologie 8: 505–688.
99. Hennig, W. 1958. Erwin Lindner zum 70. Geburtstage. Anzeiger für Schädlingskunde 31: 76–77.
100. Hennig, W. 1959. *Encarsiocera pennipes* Czerny, eine angebliche Psilide, synonym zu *Anaphalantus pennatus* Loew (Diptera: Muscidae, Coenosiini). Beiträge zur Entomologie 9: 372.
101. Hennig, W. 1959. Taschenbuch der Zoologie, Teil 2: Wirbellose II (Gliedertiere). 1st ed., 170 pp., Georg Thieme, Leipzig.
103. Hennig, W. 1959. Pyrgotidae de Madagascar (Diptera). Mémoires de l'Institut scientifique de Madagascar Série E 11: 321–353.
104. Piekarski, G.; Hennig, W. & Sibbing, B. 1959: Die geographische Verbreitung der Leishmaniasen und der Phlebotomen als ihrer Überträger in Asien (1900–1957). Text III/83–86, maps 101–120/I in Rodenwaldt, E. (ed.) Weltseuchenatlas/World Atlas of Epidemic Diseases. Falk, Hamburg.

102. Hennig, W. 1960. F.I. van Emden †. Zoologischer Anzeiger Supplement 23 (Verhandlungen der Deutschen Zoologischen Gesellschaft 1959): 528–529.

105. Hennig, W. 1960. Diptera (Brachycera): Sepsidae. Pp. 377–380 in South African Animal Life (Results of the Lund University Expedition in 1950–1951) vol. 7. Almqvist & Wiksell, Uppsala.

106. Hennig, W. 1960. Die Dipteren-Fauna von Neuseeland als systematisches und tiergeographisches Problem. Beiträge zur Entomologie 10: 221–329.

107. Hennig, W. 1961. Bericht über die Untersuchung der Typen einiger der von Rondani beschriebenen Arten aus der Familie Muscidae. Beiträge zur Entomologie 11: 225–229.

108. Hennig, W. 1962. Veränderungen am phylogenetischen System der Insekten seit 1953. Bericht über die 9. Wanderversammlung Deutscher Entomologen 1961 in Berlin. Deutsche Akademie für Landwirtschaftswissenschaften Berlin, Tagungsbericht 45: 29–42.

109. Hennig, W. 1963. Zwei neue paläarktische Arten aus der Familie Muscidae (Dipt.). Stuttgarter Beiträge zur Naturkunde Nr. 101 (Festschrift Lindner): 1–3.

110. Hennig, W. 1963. Eine neue Art der Gattung *Trichopticoides* aus Zentralasien (Diptera: Muscidae). Mitteilungen der Deutschen Entomologischen Gesellschaft 22: 55–57.

96a. Hennig, W. 1963. Taschenbuch der Zoologie, Wirbellose I, ausgenommen Gliedertiere, 2nd ed., VII + 178 pp., Edition Leipzig; Leipzig.

111. Hennig, W. 1964. Die Dipteren-Familie Sciadoceridae im Baltischen Bernstein (Diptera: Cyclorrhapha Aschiza). Stuttgarter Beiträge zur Naturkunde Nr. 127: 1–10.

112. Hennig, W. 1964. Muscidae aus dem Pamir. Stuttgarter Beiträge zur Naturkunde Nr. 131: 1–6.

101a. Hennig, W. 1964. Taschenbuch der Zoologie, Wirbellose II (Gliedertiere). 2nd ed., 169 pp., Georg Thieme, Leipzig.

113. Hennig, W. 1964. Diptera Acalyptratae aus dem Iran. Stuttgarter Beiträge zur Naturkunde Nr. 139: 1–6.

[113a]. Hennig, W. 1964. Diskussionsbemerkung Hennig. Zoologische Anzeiger 173: 63.

114. Hennig, W. 1954–1964. Muscidae. 1110 pp. + 33 plates in Lindner, E. (ed.) Die Fliegen der paläarktischen Region, Teil 63b, several "Lieferungen" from 1954 until 1964. E. Schweizerbart'sche Verlagsbuchhandlung, Stuttgart.

115. Hennig, W. 1965. Phylogenetic Systematics. Annual Review of Entomology 10: 97–116.

116. Hennig, W. 1965. Einige allgemeine Gesichtspunkte für die phylogenetische Deutung des Flügelgeäders der Dipteren. Pp. 67–69 in Papers from the 12th International Congress of Entomology, London 1964.

117. Hennig, W. 1965. Vorarbeiten zu einem phylogenetischen System der Muscidae (Diptera: Cyclorrhapha). Stuttgarter Beiträge zur Naturkunde Nr. 141: 1–100.

118. Hennig, W. 1965. Die Acalyptratae des Baltischen Bernsteins und ihre Bedeutung für die Erforschung der phylogenetischen Entwicklung dieser Dipteren-Gruppe. Stuttgarter Beiträge zur Naturkunde Nr. 145: 1–215.

96a. Hennig, W. 1965. Taschenbuch der Zoologie, Wirbellose I, ausgenommen Gliedertiere. 2nd ed., VII + 178 pp., Georg Thieme; Leipzig.

106a. Hennig, W. 1966. The Diptera Fauna of New Zealand as a problem in systematics and biogeography. Pacific Insects Monograph 9: 1–81 [English translation of no. 106 by P. Wygodzinsky].

119. Hennig, W. 1966. *Fannia scalaris* Fabricius, eine rezente Art im Baltischen Bernstein? (Diptera: Muscidae). Stuttgarter Beiträge zur Naturkunde Serie A (Biologie) 150: 1–12.

120. Hennig, W. 1966. Dixidae aus dem Baltischen Bernstein, mit Bemerkungen über einige andere fossile Arten aus der Gruppe Culicoidea (Diptera Nematocera). Stuttgarter Beiträge zur Naturkunde Nr. 153: 1–16.

121. Hennig, W. 1966. Conopidae im Baltischen Bernstein (Diptera; Cyclorrhapha). Stuttgarter Beiträge zur Naturkunde Nr. 154: 1–24.

122. Hennig, W. 1966. Einige Bemerkungen über die Typen der von Giebel 1862 angeblich aus dem Bernstein beschriebenen Insektenarten. Stuttgarter Beiträge zur Naturkunde Nr. 162: 1–7.
123. Hennig, W. 1966. Spinnenparasiten der Familie Acroceridae im Baltischen Bernstein. Stuttgarter Beiträge zur Naturkunde Nr. 165: 1–21.
124. Hennig, W. 1966. Bombyliidae im Kopal und im Baltischen Bernstein (Diptera: Brachycera). Stuttgarter Beiträge zur Naturkunde Nr. 166: 1–20.
125. Hennig, W. 1966. Phylogenetic Systematics. 263 pp., University of Illinois Press, Urbana (English translation of no. 161 by D.D. Davis & R. Zangerl).
126. Hennig, W. 1967. Die sogenannten "niederen Brachycera" im Baltischen Bernstein (Diptera: Fam. Xylophagidae, Xylomyidae, Rhagionidae, Tabanidae). Stuttgarter Beiträge zur Naturkunde Nr. 174: 1–51.
127. Hennig, W. 1967. Neue Acalyptratae aus dem Baltischen Bernstein (Diptera: Cyclorrhapha). Stuttgarter Beiträge zur Naturkunde Nr. 175: 1–27.
128. Hennig, W. 1967. Therevidae aus dem Baltischen Bernstein mit einigen Bemerkungen über Asilidae und Bombyliidae (Diptera Brachycera). Stuttgarter Beiträge zur Naturkunde Nr. 176: 1–14.
129. Hennig, W. 1967. Muscidae. Pp. 423–424 in Illies, J. (ed.) Limnofauna europaea. Gustav Fischer, Stuttgart.
96b. Hennig, W. 1967. Taschenbuch der Zoologie, Wirbellose I, ausgenommen Gliedertiere. 3rd ed., IIX + 179 pp., Georg Thieme; Leipzig (as "Taschenbuch der Speziellen Zoologie, Teil 1, Wirbellose, ausgenommen Gliedertiere", VI+203 pp., at Harri Deutsch, Thun—Frankfurt am Main 1972).
130. Hennig, W. 1968. Ein weiterer Vertreter der Familie Acroceridae im Baltischen Bernstein (Diptera: Brachycera). Stuttgarter Beiträge zur Naturkunde Nr. 185: 1–6.
131. Hennig, W. 1968. *Holopticander*, eine neue Gattung der Lauxaniidae, mit Bemerkungen über die Gattung *Hypagoga* (Diptera: Acalyptratae). Stuttgarter Beiträge zur Naturkunde Nr. 192: 1–6.
132. Hennig, W. 1968. Kritische Bemerkungen über den Bau der Flügelwurzel bei den Dipteren und die Frage nach der Monophylie der Nematocera. Stuttgarter Beiträge zur Naturkunde Nr. 193: 1–23.
101b. Hennig, W. 1968. Taschenbuch der Zoologie, Teil 2: Wirbellose II (Gliedertiere). 3rd ed., VII + 175 pp., Georg Thieme, Leipzig (as "Taschenbuch der Speziellen Zoologie, Teil 2, Wirbellose , Gliedertiere", VIII+199 pp., at Harri Deutsch, Thun—Frankfurt am Main 1972).
133. Hennig, W. 1969. Neue Gattungen und Arten der Acalyptratae. Canadian Entomologist 101: 589–633.
134. Hennig, W. 1969. Neue oder bisher ungedeutete Arten der Gattungen *Chirosia*, *Paraprosalpia* und *Craspedochoeta* (Diptera: Anthomyiidae). Stuttgarter Beiträge zur Naturkunde Nr. 208: 1–12.
135. Hennig, W. 1969. Neue Übersicht über die aus dem Baltischen Bernstein bekannten Acalyptratae (Diptera: Cyclorhapha). Stuttgarter Beiträge zur Naturkunde Nr. 209: 1–42.
136. Hennig, W. 1969. Bernsteinfossilien. Naturwissenschaft und Medizin 6 No. 26: 10–24.
137. Hennig, W. 1969. Kritische Betrachtungen über die phylogenetische Bedeutung von Bernsteinfossilien; die Gattungen *Proplatypygus* (Diptera: Bombyliidae) und *Palaeopsylla* (Siphonaptera). Memorie della Società Entomologica Italiana 48 (volume del Centenario): 57–67.
138. Hennig, W. 1969. Die Stammesgeschichte der Insekten. 436 pp., Waldemar Kramer, Frankfurt am Main. (English edition 1981, no. 138a).
139. Hennig, W. 1970. Insektenfossilien aus der unteren Kreide II. Empididae (Diptera, Brachycera). Stuttgarter Beiträge zur Naturkunde Nr. 214: 1–12.
93a. van Emden, F. & Hennig, W. 1970. Diptera. Pp. 130–141 in Tuxen, S.L. (ed.) Taxonomist's Glossary of Genitalia in Insects, 2nd ed., Munksgaard, Copenhagen.

140. Hennig, W. 1971. Zur Situation der biologischen Systematik Pp. 7–15 in: Siewing: R. (ed.) Methoden der Phylogenetik—Symposion vom 12.-13. Feb. 1970. Erlanger Forschungen Reihe B—Naturwissenschaften 4.

[140a]. Hennig, W. 1971. [Comment]. P. 28 in Schlee, D.: Die Rekonstruktion der Phylogenese mit Hennig's Prinzip. Aufsätze und Reden der Senckenbergischen Naturforschenden Gesellschaft 20: 1–62.

141. Hennig, W. 1971. Neue Untersuchungen über die Familien der Diptera Schizophora (Diptera: Cyclorrhapha). Stuttgarter Beiträge zur Naturkunde Nr. 226: 1–76.

142. Hennig, W. 1971. Insektenfossilien aus der unteren Kreide III. Empidiformia ("Microphorinae") aus der unteren Kreide und aus dem Baltischen Bernstein; ein Vertreter der Cyclorrhapha aus der unteren Kreide. Stuttgarter Beiträge zur Naturkunde Nr. 232: 1–28.

143. Hennig, W. 1971. Die Familien Pseudopomyzidae und Milichiidae im Baltischen Bernstein (Diptera: Cyclorrhapha). Stuttgarter Beiträge zur Naturkunde Nr. 233: 1–16.

144. Hennig, W. 1972. Beiträge zur Kenntnis der rezenten und fossilen Carnidae, mit besonderer Berücksichtigung einer neuen Gattung aus Chile (Diptera: Cyclorrhapha). Stuttgarter Beiträge zur Naturkunde Nr. 240: 1–20.

145. Hennig, W. 1972. Insektenfossilien aus der unteren Kreide IV. Psychodidae (Phlebotominae), mit einer kritischen Übersicht über das phylogenetische System der Familien und die bisher beschriebenen Fossilien (Diptera). Stuttgarter Beiträge zur Naturkunde Nr. 241: 1–69.

146. Hennig, W. 1972. Eine neue Art der Rhagionidengattung *Litoleptis* aus Chile, mit Bemerkungen über Fühlerbildung und Verwandtschaftsbeziehungen einiger Brachycerenfamilien (Diptera, Brachycera). Stuttgarter Beiträge zur Naturkunde Nr. 242: 1–18.

147. Hennig, W. 1972. *Pseudomyopina facetti* (Bezzi): eine bisher falsch gedeutete Art der Gattung *Pseudomyopina* aus den Alpen. Stuttgarter Beitraege zur Naturkunde Serie A (Biologie) 247: 1–4.

148. Hennig, W. 1973. Diptera (Zweiflügler). 337 pp. in Helmcke, J.-G.; Starck, D. & Wermuth, H. (eds.) Handbuch der Zoologie Band 4, 2. Hälfte, 2. Teil, Nr. 31 (Lieferung 20). Walter de Gruyter, Berlin – New York.

149. Hennig, W. 1974. Kritische Bemerkungen zur Frage "Cladistic analysis or cladistic classification?". Zeitschrift für zoologische Systematik und Evolutionsforschung 12: 279–294.

149a. Hennig, W. 1975. "Cladistic analysis or cladistic classification?" A reply to Ernst Mayr. Systematic Zoology 24: 244–256 (translation of no. 149).

150. Hennig, W. 1975. Die systematische Stellung der Gattung *Taeniomyia* Stein und einiger anderer neotropischer Anthomyiidae (Diptera: Anthomyiidae). Stuttgarter Beiträge zur Naturkunde Serie A Nr. 281: 1–6.

151. Hennig, W. 1976. Das Hypopygium von *Lonchoptera lutea* Panzer und die phylogenetischen Verwandtschaftsbeziehungen der Cyclorrhapha (Diptera). Stuttgarter Beiträge zur Naturkunde Serie A Nr. 283: 1–63.

152. Hennig, W. 1966–1976. Anthomyiidae. 974 pp. + 114 plates in Lindner, E. (ed.) Die Fliegen der paläarktischen Region, Teil 63a, several "Lieferungen" from 1966 until 1976. E. Schweizerbart'sche Verlagsbuchhandlung, Stuttgart.

153. Hennig, W. 1976. Klaus Günther 7.10.1907 bis 1.8.1975 Verhandlungen der Deutschen zoologischen Gesellschaft 69: 297–298.

posthumous publications:

154 or 129a. Hennig, W. 1978. Muscidae. Pp. 490–491 in Illies, J. (ed.) Limnofauna europaea, 2nd ed., Gustav Fischer, Stuttgart.

158. Hennig, W. 1978. Über Muscidae (Diptera) der Kanarischen Inseln. Stuttgarter Beiträge zur Naturkunde Serie A (Biologie) 311: 1–12.

[166] Hennig, W. 1978. Die Stellung der Systematik in der Zoologie. Entomologica Germanica 4: 193–199.
155. Bachmayer, F.; Schulz, E.; Hennig, W. & Maraspin, D. 1978. Ein bemerkenswerter Insektenrest im fossilen Harz des Glaukonitsandsteines (Eggenburgien) der Aufschlußbohrung "Herzogbirbaum 1" (Niederösterreich). Annalen des Naturhistorischen Museums in Wien 81: 113–120.
162. Hennig, W. & Schlee, D. 1978. Abriß der phylogenetischen Systematik. Stuttgarter Beiträge zur Naturkunde Serie A (Biologie) 319: 1–11.
157 or 96c. Hennig, W. 1979. Taschenbuch der Speziellen Zoologie Teil 1. Wirbellose 1 (ausgenommen Gliedertiere) (Hennig, Wolfgang ed.). 4th ed., 392 pp., Gustav Fischer, Jena.
138a. Hennig, W. 1981. Insect Phylogeny, translated and edited by A.C. Pont, with revisionary notes by D. Schlee. XXII + 514 pp., John Wiley & Sons, Chichester etc.
161. Hennig, W. 1982. Phylogenetische Systematik. 246 pp., Paul Parey, Berlint – Hamburg.
160. Hennig, W. 1983. Stammesgeschichte der Chordaten (Fortschritte in der zoologischen Systematik und Evolutionsforschung 2). 108 pp., Paul Parey, Hamburgt – Berlin.
[167] Hennig, W. 1984. Aufgaben und Probleme stammesgeschichtlicher Forschung. 65 pp., Paul Parey, Berlin – Hamburg.
96d. Hennig, W. 1984. Taschenbuch der Speziellen Zoologie Teil 2 Wirbellose 1 (ausgenommen Gliedertiere) (Hennig, Wolfgang ed.). 5th ed., 392 S., Gustav Fischer, Jena.
101c. Hennig, W. 1986. Taschenbuch der Speziellen Zoologie Teil 2 Wirbellose 2 (Gliedertiere) (Hennig, W. & Mickoleit, G. ed.). 4th ed., 335 S., Gustav Fischer, Jena.
96e. Hennig, W. 1994. Taschenbuch der speziellen Zoologie (Hennig, W. & Mickoleit, G., eds.) Teil 1—Wirbellose 1 (ausgenommen Gliedertiere). 6. Auflage. 392 pp., Gustav Fischer, Jenat – Stuttgart – New York.
101d. Hennig, W. 1994. Taschenbuch der speziellen Zoologie (Hennig, W. & Mickoleit, G., eds.) Teil 1—Wirbellose 2 (Gliedertiere). 5. Auflage. 335 pp., Gustav Fischer, Jenat – Stuttgartt – New York.

2.2. *Publications by Other Authors*

Aly, G. 2006. Hitlers Volksstaat: Raub, Rassenkrieg und nationaler Sozialismus. 480 pp., S. Fischer, Frankfurt am Main.
Amlung, U. 2009. Landesschule Dresden—Entstehung, Konzeption und Praxis (1920–1933). Pp. 24–49 in: Deutsche Gesetzliche Unfallversicherung e.V.(ed.) Lernräume—von der Landesschule Dresden zur Akademie. Sandstein Verlag, Dresden.
Anonymous 1956. Zum 70. Geburtstag von Paul Buchner. Zeitschrift für Morphologie und Ökologie der Tiere 44(6): II–III.
——. 1978. In memoriam: Willi Hennig (20.4.1913–5.11.1976). Beiträge zur Entomologie 28: 169–177.
Ashlock, P.D. 1971. Monophyly and associated terms. Systematic Zoology 20: 63–69.
Ax, P. 1977a. Professor Dr.Dr.h.c. Willi Hennig †. Zoomorphologie 87: 1–2.
——. 1977b. Willi Hennig 20.4.1913 bis 5.11.1976. Verhandlungen der Deutschen zoologischen Gesellschaft 70: 346–347.
——. 1984. Das Phylogenetische System. 349 pp., Gustav Fischer, Stuttgart und New York.
Baroni-Urbani, C. 1990. Search for the evolutionary roots of cladistics (a simplified conspectus of hologenetic theory). Pp. 1–7. Osaka Group for the Study of Dynamic Structures (newsletter), November.
Bechly, G. 2010. Additions to the fossil dragonfly fauna of the Lower Cretaceous Crato Formation of Brazil (Insecta: Odonata). Palaeodiversity 3 Supplement ("Contributions to the Willi-Hennig-Symposium on Phylogenetics and Evolution, University of Hohenheim, 29 September–2 October 2009"): 11–77.
Bernier, R. 1984. The species as an individual: facing essentialism. Systematic Zoology 33: 460–469.

Bock, W.J. 1968. Phylogenetic systematics, cladistics and evolution. Evolution 22: 646–648.
Bremer, K. & Wanntorp, H.-E. 1978. Phylogenetic systematics in botany. Taxon 27: 317–329.
Brundin, L. 1965. On the real nature of transantarctic relationships. Evolution 19: 496–505.
Brundin, L.Z. 1966. Transantarctic relationships and their significance as evidenced by chironomid midges. With a monograph of the subfamilies Podonominae and Aphroteniinae and the austral Heptagyiae (Kungliga Svenska vetenskapsakademiens handlingar—4. serie, band 11, nr. 1). 472 pp., Almqvist & Wiksell, Stockholm.
Bunge, M. 1981. Biopopulations, not biospecies, are individuals and evolve. Behavioral and Brain Sciences 4: 284–285.
Byers, G.W. 1977. In memoriam Willi Hennig (1913–1976). Journal of the Kansas entomological Society 50: 272–274.
Cain, A.J. & Harrison, G.A. 1960. Phyletic weighting. Proceedings of the Zoological Society of London 135: 1–31.
Colless, D.H. 1977. In memoriam Willi Hennig. Journal of the Australian Entomological Society 16: 6.
Croizat-Chaley, L. 1978. Hennig (1966) entre Rosa (1918) y Lovtrup (1977): medio siglo de "sistema filogenetica". Boletino de la Academia de Ciencias Fisicas, Matematicas y Naturales 38(116): 59–146.
Cumming, J.M.; Sinclair, B.J.; Brooks, S.E.; O'Hara, J.E. & Skevington, J.H. 2011. The history of dipterology at the Canadian National Collection of Insects, with special reference to the Manual of Nearctic Diptera. Canadian Entomologist 143: 539–577.
Danser, B.H. 1942. Typologische und phylogenetische Systematik. Physis 1: 52–63.
De Queiroz, K. & Gauthier, J. 1992. Phylogenetic taxonomy. Annual Review of Ecology and Systematics 23: 449–480.
de Sousa, R. 1989. Kinds of kinds: individuality and biological species. International Studies in the Philosophy of Science 3: 119–135.
Dupuis, C. 1979. La "systématique phylogénétique" de W. Hennig (Historique, discussion, choix de références). Cahiers des Naturalistes. Bulletin des Naturalistes Parisiens Nouvelle Série 34: 1–69 (1978).
——. 1984. Willi Hennig's impact on taxonomic thought. Annual Review of Ecology and Systematics 15: 1–24.
——. 1990. Hennig, Emil Hans Willi. Pp. 407–410 in Holmes, F.L. (ed.) Dictionary of Scientific Biography 17, Supplement 2. Charles Scribner's Sons, New York.
Ebert, W.; Rohlfien, K.; Petersen, G. & Friese, G. 1986. Einhundert Jahre Deutsches Entomologisches Institut. Beiträge zur Entomologie 36: 5–62.
Engel, M.S. & Kristensen, N.P. 2013. A history of entomological classification. Annual Review of Entomology 58: 585-607.
Evenhuis, N.L. 2004. Biography of D. Elmo Hardy (1914–2002). Pp. 1–10 in: Evenhuis, N.L. & Kaneshiro, K.Y. (eds.) D. Elmo Hardy Memorial Volume—Contributions to the Systematics and Evolution of Diptera (Bishop Museum Bulletin in Entomology 12) Bishop Museum, Honolulu.
Farris, J.S. 1970. Methods for computing Wagner trees. Systematic Zoology 19: 83–92.
——. 1985. The pattern of cladistics. Cladistics 1: 190–201.
——. 1988. Hennig86 version 1.5, Computer programme and manual, published by the author.
——. 2012. Early Wagner trees and the "cladistic redux". Cladistics 28: 545–547.
Fischer, C.; Herzog, E.; Hoffmann, W.; Radecker, F.; Thieme, E. & Thomas, C. 1927. Zur Einweihung der Landesschule Dresden in Klotzsche am 15. Oktober 1927. Festschrift der Landesschule, 93 pp., J. Päßler, Dresden.
Forey, P.L.; Humphries, C.J.; Kitching, I.J.; Scotland, R.W.; Siebert, D.J. & Williams, D.M. 1992. Cladistics. A Practical Course in Systematics (Systematics Association Publications 10). XI + 191 pp., Clarendon Press, Oxford.
Franz, V. 1943. Die Geschichte der Tiere. Pp. 219–296 in: Heberer, G. (ed.) Die Evolution der Organismen Gustav Fischer, Jena.

Gersch, M. 1979. Paul Buchner (12.4.1886–19.10.1978). Verhandlungen der Deutschen Zoologischen Gesellschaft 72: 319–321.
Ghiselin, M.T. 1987. Response to commentary on the individuality of species. Biology & Philosophy 2: 207–212.
——. 1989. Sex and the individuality of species: A reply to Mishler and Brandon. Biology & Philosophy 4: 73–76.
Gilmour, J.S.L. 1940. Taxonomy and philosophy. Pp. 461–474 in: Huxley, J. (ed.) The New Systematics. Oxford University Press, London.
Griffiths, G.C.D. 1974. On the foundation of biological systematics. Acta Biotheoretica 23(3–4): 85–131.
Günther, K. 1950. Ökologische und funktionelle Anmerkungen zur Frage des Nahrungserwerbs bei Tiefseefischen mit einem Exkurs über die ökologischen Zonen und Nischen. Pp. 55–93 in: Grüneberg, H. & Ulrich, W. (eds.) Moderne Biologie. Festschrift zum 60. Geburtstag von Hans Nachtsheim. F.W. Peters, Berlin.
——. 1956. Systematik und Stammesgeschichte der Tiere 1939–1953. Fortschritte der Zoologie Neue Folge 10: 33–278.
——. 1962. Systematik und Stammesgeschichte der Tiere 1954–1959. Fortschritte der Zoologie Neue Folge 14: 268–547.
——. 1971. Natürliches oder phylogenetisches System. Pp. 29–41 in: Siewing, Rolf (ed.) Methoden der Phylogenetik. Symposion vom 12.–13. Februar 1970. Erlanger Forschungen Reihe B—Naturwissenschaften 4.
Hadži, J. 1953. An attempt to reconstruct the system of animal classification. Systematic Zoology 2(4): 145–154.
Haeckel, E. 1866. Generelle Morphologie der Organismen. Allgemeine Grundzüge der organischen Formenwissenschaft, mechanisch begründet durch die von Charles Darwin reformierte Descendenz-Theorie. Vol. 1 XXXII + 574 pp. + 2 plates, vol. 2 CLX + 462 pp. + 8 plates. Georg Reimer, Berlin.
Haffer, J. 2003. Wilhelm Meise (1901–2002), ein führender Ornithologe Deutschlands im 20. Jahrhundert. Verhandlungen des Naturwissenschaftlichen Vereins in Hamburg Neue Folge 40: 117–140.
Herter, K. & Strübing, H. 1975. Klaus Günther † (7.10.1907–1.8.1975). Sitzungsberichte der Gesellschaft Naturforschender Freunde zu Berlin Neue Folge, Beiheft zu Band 15: 1–11.
Hoerschelmann, H. & Neumann, J. 2003. Prof. Dr. Wilhelm Meise 12.9.1901–24.8.2002. Journal of Ornithology 144: 110–111.
Hull, D.L. 1967. Certainty and circularity in evolutionary taxonomy. Evolution 21: 174–189.
——. 1976. Are species really individuals? Systematic Zoology 25: 174–191.
——. 1988. Science as a Process. XIII + 586 pp., University of Chicago Press, Chicago – London.
——. 1989. The evolution of phylogenetic systematics. Pp. 3–15 in: Fernholm, B.; Bremer, K. & Jörnvall, H. (eds.) The Hierarchy of Life. Elsevier (Excerpta Medica), Amsterdam etc.
Hutterer, R. 1998. Blücherhof in Mecklenburg: Mustergut, Sommersitz und Gästehaus Alexander Koenigs von 1904–1940. Tier und Museum 6: 3–18.
Huxley, J.S. 1958. Evolutionary processes and taxonomy with special reference to grades. Pp. 21–39 in: Hedberg, O. (ed.) Systematics Today, Uppsala Universitets Arsskrift 1958(6) Uppsala Universitet, Uppsala.
Jahn, I. 1992. Geschichte der Systematik. Pp. 329–333 in Schmitt, M. (ed.) Lexikon der Biologie vol. 10 (Biologie im Überblick). Herder Verlag, Freiburg im Breisgau.
Jefferies, R.P.S. 1979. (17) The origin of chordates—a methodological essay. Pp. 443–447 in: House, M.R. (ed.) The Origin of Major Invertebrate Groups (Systematics Association Special Volume No. 12). Academic Press, London – New York.
Kämpfe, L. & Michalik, P. 2011. 175 Jahre Zoologie in Greifswald—Ein Rückblick. Greifswalder Universitätsreden Neue Folge 143: 7–27.
Kiriakoff, S.G. 1977. Necrologie: Willi Hennig (1913–1976). Bulletin et Annales de la Société Royale d'Entomologie Belgique 113: 240–243.

Kitching, I.J., Forey, P.L., Humphries, C.J. & Williams, D.M. 1998. Cladistics. The Theory and Practice of Parsimony Analysis, 2nd ed. Oxford University Press, Oxford.

Klietmann, K.-G. 1996. Auszeichnungen des Deutschen Reiches 1936–1945. Eine Dokumentation ziviler und militärischer Verdienst- und Ehrenzeichen. Motorbuch-Verlag, Stuttgart.

Kluge, A.G. 2009. Explanation and falsification in phylogenetic inference: Exercise in Popperian philosophy. Acta biotheoretica 57: 171–186.

Kluge, A.G. & Farris, J.S. 1969. Quantitative phyletics and the evolution of anurans. Systematic Zoology 18: 1–32.

Kohring, R. 1991. Walter Georg Kühne, 1911–1991. News Bulletin Society of Vertebrate Paleontology 153: 46–47.

Kohring, R. & Schlüter, T. 1997. Einleitung zu: Paläontologische Essays 1943–1990 von Walter Georg Kühne. Documenta naturae 113: 3–24.

Königsmann, E. 1975. Termini der phylogenetischen Systematik. Biologische Rundschau 13: 99–115.

Kraus, O. 1984. Die Veranstaltung "Phylogenetisches Symposion": Rückblick auf 25 Tagungen (1955–1982). Verhandlungen des Naturwissenschaftlichen Vereins in Hamburg Neue Folge 27: 277–289.

Kraus, O. & Hoßfeld, U. 1998. 40 Jahre "Phylogenetisches Symposium" 1956–1997, eine Übersicht—Anfänge, Entwicklung, Dokumentation und Wirkung. Jahrbuch für Geschichte und Theorie der Biologie 5: 157–186.

Kuhn, T.S. 1962. The structure of scientific revolutions. XV + 172 pp., University of Chicago Press, Chicago IL.

Kühne, W.G. 1978. Willi Hennig 1913–1976: Die Schaffung einer Wissenschaftstheorie. Entomologica Germanica 4: 374–376.

Kühnelt, W. 1962. Prinzipien der Systematik. Pp. 1–16 in Handbuch der Biologie Bd. 6, Teil 2/1. Akademische Verlagsgesellschaft Athenaion, Konstanz.

Lauterbach, K.-E. 1989. Das Pan-Monophylum—Ein Hilfsmittel für die Praxis der Phylogenetischen Systematik. Zoologischer Anzeiger 223: 139–156.

Lorenzen, S. 1994. Phylogenetische Systematik gestern, heute und morgen. Biologie in unserer Zeit 24(4): 200–206.

Löther, R. 1972a. Die Beherrschung der Mannigfaltigkeit. 285 pp., Gustav Fischer, Jena.

——. 1972b. Biologie und Weltanschauung, eine Einführung in philosophische Probleme der Biologie vom Standpunkt des dialektischen und historischen Materialismus. 128 pp., Urania, Leipzig.

Malmeström, E. & Uggla, A.Hj. 1957. Vita Caroli Linnaei—Carl von Linnés Självbiografier. 235 pp., Almquist & Wiksell, Stockholm.

Mayer, G. 1988. Württembergische Paläontologen II.—Nachträge, Ergänzungen und Berichtigungen. Jahresheft der Gesellschaft für Naturkunde in Württemberg 143: 111–147.

Mayr, E. 1942. Systematics and the origin of species. From the viewpoint of a zoologist. XIV + 334 pp., Columbia Univ. Press, New York.

——. 1965. Numerical phenetics and taxonomic theory. Systematic Zoology 14: 73–97.

——. 1969. Principles of Systematic Zoology. XI + 428 pp., McGraw-Hill, New York etc.

——. 1974. Cladistic analysis or cladistic classification. Zeitschrift für zoologische Systematik und Evolutionsforschung 12: 94–128.

——. 1984. Die Entwicklung der biologischen Gedankenwelt. XXI + 766 pp., Springer, Berlin etc. (Orig. The Growth of Biological Thought, 1982).

Meier, R. 2005. Role of dipterology in phylogenetic systematics: The insight of Willi Hennig. Pp. 45–62 in: Yeates, D.K. & Wiegmann, B.M. (eds.) The Evolutionary Biology of Flies. Columbia University Press, New York.

Mühlmann, W.E. 1939. Geschichtliche Bedingungen, Methoden und Aufgaben der Völkerkunde. Pp. 1–43 in Preuß, K. Th. & Thurnwald, R. (eds.) Lehrbuch der Völkerkunde. 2nd ed., Enke, Stuttgart.

Naef, A. 1917. Die individuelle Entwicklung organischer Formen als Urkunde ihrer Stammesgeschichte. 77 pp., Gustav Fischer, Jena.

——. 1919. Idealistische Morphologie und Phylogenetik (Zur Methodik der systematischen Morphologie). VI + 77 pp., Gustav Fischer, Jena.

Nagel, E. 1961. The structure of science. Problems in the logic of scientific explanation. XIII + 618 pp., Harcourt, Brace & World, New York – Burlingame.

Nelson, G.J. 1971. 'Cladism' as a philosophy of classification. Systematic Zoology 20: 373–376.

——. 1972. Phylogenetic relationship and classification. Systematic Zoology 21: 227–231.

——. 1979. Cladistic analysis and synthesis: Principles and definitions, with a historical note on Adanson's Familles des Plantes 1763–1764). Systematic Zoology 28: 1–21.

Oldroyd, H. 1958. Obituary—I.F. van Emden. Entomologist's monthly Magazine 94: 228–229 + 1 photo.

Osche, G. 1966. Grundzüge der allgemeinen Phylogenetik. Pp. 817–906 in: Handbuch der Biologie 3/2. Akademische Verlagsgeellschaft Athenaion, Frankfurt am Main.

Patterson, C. 1980. Cladistics. Biologist 27: 234–240.

Peters, G. 1971a. Probleme der Taxonomie im Biologieunterricht (I). Biologie in der Schule 20(10): 401–412.

——. 1971b. Probleme der Taxonomie im Biologieunterricht (II). Biologie in der Schule 20(12): 505–519.

Peters, G. & Klausnitzer, B. 1978. Phylogenetische Systematik als Methode zur Erforschung der Stammesgeschichte der Tiere. Biologische Rundschau 16: 8–98.

Plate, L. 1922. Allgemeine Zoologie und Abstammungslehre, Erster Teil. VI + 629 pp., Gustav Fischer, Jena.

Platnick, N.I. 1977. Cladograms, phylogenetic trees, and hypothesis testing. Systematic Zoology 26: 438–442.

——. 1979. Philosophy and the transformation of cladistics. Systematic Zoology 28: 537–546.

——. 1989. Cladistics and phylogenetic analysis today. Pp. 17–24 in: Fernholm, B.; Bremer, K. & Jörnvall, H. (eds.) The Hierarchy of Life. Elsevier (Excerpta Medica), Amsterdam etc.

Platnick, N.I. & Nelson, G.J. 1988. Spanning-tree biogeography: Shortcut, detour, or dead end? Systematic Zoology 37: 410–419.

Pont, A.C. 1981. Foreword. Pp. IX–XI in Hennig, W. Insect Phylogeny. John Wiley & Sons, Chichester etc.

Popper, K.R. 1973. Objektive Erkenntnis. 417 pp., Hoffmann & Campe, Hamburg (Orig. Objective Knowledge, 1972).

Remane, A. 1952. Die Grundlagen des natürlichen Systems, der vergleichenden Anatomie und der Phylogenetik. Theoretische Morphologie und Systematik I. VI + 400 pp., Akademische Verlagsgesellschaft Geest & Portig, Leipzig.

Rensch, B. 1947. Neuere Probleme der Abstammungslehre. VII + 407 pp., Ferdinand Enke Verlag, Stuttgart.

Richter, S. & Meier, R. 1994. The development of phylogenetic concepts in Hennig's early theoretical publications (1947–1966). Systematic Biology 43: 212–221.

Rieppel, O. 2007a. The metaphysics of Hennig's phylogenetic systematics: substance, events and laws of nature. Systematics and Biodiversity 5: 345–360.

——. 2007b. Species: kinds of individuals or individuals of a kind. Cladistics 23: 373–384.

——. 2010. Reydon on species, individuals and kinds: a reply. Cladistics 26: 341–343.

——. 2011a. The dark side of the moon. Cladistics 27: 34–40.

——. 2011b. Willi Hennig's dichotomization of nature. Cladistics 27: 103–112.

Rosa, D. 1903. Die Progressive Reduktion der Variabilität und ihre Beziehungen zum Aussterben und zur Entstehung der Arten. III + 106 pp., Gustav Fischer, Jena (Ital. Orig.: La riduzione progressiva della variabilità. C. Clausen, Torino 1899).

——. 1923. Qu'est-ce que l'hologénèse? Scientia—Rivista di Scienza 33: 113–124.

——. 1988. The theory of hologenesis 1909–1918). Rivista di Biologia—Biology Forum 81: 613–615.

Satzinger, H. 2009. Differenz und Vererbung: Geschlechterordnungen in der Genetik und Hormonforschung 1890–1950. 484 pp., Böhlau, Köln etc.

Schlee, D. 1971. Die Rekonstruktion der Phylogenese mit Hennig's Prinzip. Aufsätze und Reden der Senckenbergischen Naturforschenden Gesellschaft 20: 1–62. Waldemar Kramer, Frankfurt/Main.

——. 1977. Willi Hennig. Jahresheft der Gesellschaft für Naturkunde in Württemberg 132: 196–197.

——. 1978. In memoriam Willi Hennig 1913–1976—eine biographische Skizze. Entomologica Germanica 4: 377–391.

——. 1981. Introduction to the English Edition. Pp. XIII–XV in Hennig, Willi: Insect Phylogeny. John Wiley & Sons, Chichester etc.

Schlüter, T. 1981. Walter Georg Kühne 70 Jahre alt—eine biographische Skizze. Berliner geowissenschaftliche Abhandlungen Reihe A 32: 4–17 + 1 plate.

Schmitt, M. 1985. Thesen zur Theorie und Methodologie der Phylogenetischen Systematik. Zoologische Beiträge Neue Folge 29: 19–33.

——. 1987a. Zoologische Systematik und Evolutionsökologie—über Klaus Günthers wissenschaftliches Werk. Sitzungsberichte der Gesellschaft Naturforschender Freunde zu Berlin Neue Folge 27: 165–182.

——. 1987b. 'Ecological niche' sensu Günther and 'ecological licence' sensu Osche—two valuable but poorly appreciated explanatory concepts. Zoologische Beiträge Neue Folge 31: 49–60.

——. 1996. Klaus Günthers Bedeutung für die Phylogenetische Systematik. Sitzungsberichte der Gesellschaft Naturforschender Freunde zu Berlin Neue Folge 35: 13–25.

——. 1997. Willi Hennig's handschriftliche Urfassung der "Grundzüge einer Theorie der Phylogenetischen Systematik". Pp. 42–43 in: von Campenhausen, C. (ed.) Andenken zoologischer Vergangenheit. Katalog zur Ausstellung im Naturhistorischen Museum Mainz (20. Mai–06. Juli 1997).

——. 2001. Willi Hennig (1913–1976). Pp. 316–343, 541–546 in Jahn, I. & Schmitt, M. (eds.) Darwin and Co., eine Geschichte der Biologie in Portraits vol. 2. C.H. Beck, München.

——. 2002. Willi Hennig (1913–1976) als akademischer Lehrer. Pp. 53–64 in Schulz, J. (ed.) Fokus Biologiegeschichte. Zum 80. Geburtstag der Biologiehistorikerin Ilse Jahn. Akadras, Berlin.

——. 2003. Willi Hennig and the rise of cladistics. Pp. 369–379 in Legakis, A.; Sfenthourakis, S.; Polymeni, R. & Thessalou-Legaki, M. (eds.) The New Panorama of Animal Evolution (Proc. 18th Int.Congr.Zoology). Pensoft, Sofia – Moscow.

——. 2004. "Willi Hennig". Pp. 186–187 in Hoffmann, D.; Laitko, H. & Müller-Wille, S. (eds.) Lexikon der bedeutenden Naturwissenschaftler. Spektrum Akademischer Verlag, Heidelberg.

——. 2004. "Klaus Günther". P. 132 in Hoffmann, D.; Laitko, H. & Müller-Wille, S. (eds.) Lexikon der bedeutenden Naturwissenschaftler. Spektrum Akademischer Verlag, Heidelberg [Günther's place of birth is incorrectly given here as "Cottbus"].

——. 2008. Carl Linnaeus, the order of nature, and binominal names. Deutsche entomologische Zeitschrift 55: 13–17.

——. 2010a. Willi Hennig, the cautious revolutioniser. Paleodiversity 3 Supplement ("Contributions to the Willi-Hennig-Symposium on Phylogenetics and Evolution, University of Hohenheim, 29 September–2 October 2009"): 3–9.

——. 2010b. Rolf Löther und die "linke" Biologie an der Freien Universität Berlin in den 1970er Jahren. Pp. 107–114 in: Jahn, I. & Wessel, A. (eds.) Für eine Philosophie der Biologie/ For a Philosophy of Biology. Festschrift to the 75th Birthday of Rolf Löther. Kleine Verlag, München.

——. in press. Willi Hennig's part in the history of systematics. In: Hamilton, A. (ed.) Patterns in Nature: Historical and Conceptual Foundations of Phylogenetic Systematics. University of California Press, Berkeley.

Schmitt, M. & Misof, B. 2009. (8) Phylogenetik. Pp. 315–351 in: Munk, K. (ed.) Taschenlehrbuch Biologie: Ökologie—Evolution. Georg Thieme Verlag, Stuttgart.
Schmitt, M. & Speck, T. 1992. Anliegen und Methode der biologischen Systematik. Pp. 1–7 in: Schmitt, M. (ed.) Lexikon der Biologie, Band 10 (Biologie im Überblick). Herder, Freiburg – Basel – Wien.
[Schüz, E.] 1969. Staatliches Museum für Naturkunde in Stuttgart 1967 und 1968, A. Bericht des Direktors für 1968. Jahresheft der Gesellschaft für Naturkunde in Württemberg 124: 19–21.
Schwanitz, F. 1943. Genetik und Evolutionsforschung bei Pflanzen. Pp. 430–478 in: Heberer, G. (ed.) Die Evolution der Organismen Gustav Fischer, Jena.
Senglaub, K. 2001. Erinnerungen eines ehemaligen Zoologie-Studenten an seinen akademischen Lehrer Willi Hennig im Winter-Semester 1946/47 in Leipzig. Sitzungsberichte der Gesellschaft naturforschender Freunde Berlin Neue Folge 39: 165–168 (2000).
Simpson, G.G. 1944. Tempo and Mode in Evolution. XI + 331 pp., Columbia University Press, New York.
Sokal, R.R. & Sneath, P.H.A. 1963. Principles of Numerical Taxonomy. XVI + 359 pp., Freeman, San Francisco.
Stamos, D.N. 1998. Buffon, Darwin, and the non-individuality of species—a reply to Jean Gayon. Biology & Philosophy 13: 443–470.
Stammer, H.-J. 1946. Zum 60. Geburtstag Paul Buchners. Die Naturwissenschaften 33(3): 96.
Steffan, A.W. 1962. Professor Dr. Karl Strenzke †. Anzeiger für Schädlingskunde 35(5): 74–75.
——. 1976. Fabricius-Preisträger 1975: Professor Dr. Fritz Peus, Berlin. Entomologica Germanica 2: 388–393.
Steinbacher, J. 1972. Dr. phil. habil. Wilhelm Meise zum 70. Geburtstag. Abhandlungen und Verhandlungen des naturwissenschaftlichen Vereins in Hamburg Neue Folge 16: 7–18.
Sudhaus, W. 1996. Die Bedeutung von Klaus Günther für die Evolutionsbiologie. Sitzungsberichte der Gesellschaft Naturforschender Freunde zu Berlin Neue Folge 35: 27–55.
——. 2007. Die Notwendigkeit morphologischer Analysen zur Rekonstruktion der Stammesgeschichte. Species, Phylogeny and Evolution 1: 17–32.
Sudhaus, W. & Rehfeld, K. 1992. Einführung in die Phylogenetik und Systematik. XI + 241 pp., Gustav Fischer, Stuttgart.
Sulloway, F.J. 1996. Born to Rebel: Birth Order, Family Dynamics, and Creative Lives. 654 pp. Pantheon, New York.
Throckmorton, L.H. 1968. Concordance and discordance of taxonomic characters in *Drosophila* classification. Systematic Zoology 17(4): 355–387.
Urich, K. 1975. Klaus Günther † (7.X.1907–1.VIII.1975). Zoologische Beiträge Neue Folge 21: 347–361.
Vogel, J. & Xylander, W.E.R. 1999. Willi Hennig—Ein Oberlausitzer Naturforscher mit Weltgeltung. Berichte der Naturforschenden Gesellschaft der Oberlausitz 7/8: 145–155.
von Wahlert, G. 1961. Die Entstehung der Plattfische durch ökologischen Funktionswechsel. Zoologische Jahrbücher Systematik 89: 1–42.
——. 1973. Phylogenie als ökologischer Prozeß. Naturwissenschaftliche Rundschau 26: 247–254.
Wägele, J.-W. 2000. Grundlagen der Phylogenetischen Systematik. 315 pp., Dr. Friedrich Pfeil, München.
——. 2004. (5) Hennig's Phylogenetic Systematics brought up to date. Pp. 101–125 in: Williams, D.M. & Forey, P.L. (eds.) Milestones in Systematics. CRC Press, London etc.
Wagner, W.H. jr. 1961. Problems in the classification of ferns. Pp. 841–844 in: Recent Advances in Botany (From Lectures and Symposia presented to the 9th International Botanical Congress Montreal 1959), Vol 1. University of Toronto Press, Toronto.
Watrous, L.E. & Wheeler, Q.D. 1981. The out-group comparison method of character analysis. Systematic Zoology 30: 1–11.

Wheeler, Q.D. 2008. Undisciplined thinking: morphology and Hennig's unfinished revolution. Systematic Entomology 33: 2–7.
Wickler, W. 1967. Vergleichende Verhaltensforschung und Phylogenetik. Pp. 420–508 in: Heberer, G. (ed.) Die Evolution der Organismen Gustav Fischer, Stuttgart.
Wiesemüller, B.; Rothe, H. & Henke, W. 2003. Phylogenetische Systematik. X + 189 pp., Springer; Berlin etc.
Wiley, E.O. 1980. Is the evolutionary species fiction?—A consideration of classes, individuals and historical entities. Systematic Zoology 29: 76–80.
——. 1981. Phylogenetics. The Theory and Practice of Phylogenetic Systematics. XV + 439 pp., John Wiley & Sons, New York etc.
Wiley, E.O. & Lieberman, B.S. 2011. Phylogenetics. The Theory and Practice of Phylogenetic Systematics. 2nd ed., 424 pp., John Wiley & Sons, New York etc.
Willmann, R. 1989. Paleontology and the systematization of natural taxa. Abhandlungen des Naturwissenschaftlichen Vereins in Hamburg Neue Folge 28: 267–291.
Windelband, W. 1904. Geschichte und Naturwissenschaft, Rektoratsrede der Universität Strassburg 1894. 3rd. ed., 27 pp., Heitz & Mündel, Strassburg.
Wood, D.M. 1989. Preface. Pp. V–VI in: McAlpine, J.F. & Wood, D.M. (eds.) Manual of Nearctic Diptera volume 3. Research Branch Agriculture Canada, Ontario.
Yoon, C.K. 2009. Naming Nature—The Clash Between Instinct and Science. VIII + 344 pp., W.W. Norton & Company, New York – London.
Ziegler, B. 1991. Chronologie. Stuttgarter Beiträge zur Naturkunde Serie C 30: 2–4.
Zimmermann, W. 1937. Arbeitsweise der botanischen Phylogenetik und anderer Gruppierungswissenschaften. Pp. 941–1053 in Abderhalden, E. (ed.) Handbuch der biologischen Arbeitsmethoden Abt. 9, Teil 3, 2. Hälfte. Urban & Schwarzenberg, Berlin – Wien.
——. 1943. Die Methoden der Phylogenetik. Pp. 20–56 in: Heberer, G. (ed.) Die Evolution der Organismen Gustav Fischer, Jena.
——. 1953. Evolution—die Geschichte ihrer Probleme und Erkenntnisse. IX + 623 pp., Carl Alber, Freiburg – München.
——. 1967. Methoden der Evolutionswissenschaft (=Phylogenetik). Pp. 61–160 in: Heberer, G. (ed.) Die Evolution der Organismen. 3. Auflage, 1. Band. Gustav Fischer, Stuttgart.

NAME INDEX

(biographical dates of professors at Leipzig University from the catalogus professorum lipsiensis (Doc. 016), if not stated otherwise)

INDEX OF SUBJECTS

www.ingramcontent.com/pod-product-compliance
Lightning Source LLC
Chambersburg PA
CBHW071226230926
51849CB00001B/38